CW01558962

A GRANTA lead title

# THE GHOST
# OF THE
# MOUNTAINS

# THE GHOST
# OF THE
# MOUNTAINS

*Unravelling the Secrets
of the Snow Leopard*

## KULBHUSHANSINGH
## SURYAWANSHI

GRANTA

Granta Publications, 12 Addison Avenue, London W11 4QR
First published in Great Britain by Granta Books, 2026

A CIP catalogue record for this book
is available from the British Library.

1 3 5 7 9 10 8 6 4 2

ISBN 978 1 80351 184 9 (hardback)
ISBN 978 1 80351 185 6 (ebook)

Typeset in Caslon by M Rules
Printed and bound by CPI Group (UK) Ltd, Croydon, CR0 4YY

www.granta.com

The manufacturer's authorised representative in the EU
for product safety is Authorised Rep Compliance Ltd,
71 Lower Baggot Street, Dublin D02 P593, Ireland
www.arccompliance.com

*To my mother Sunita,*
*sister Neha,*
*wife Bhagya,*
*daughter Tara,*
*and my father Tatya*

# Contents

<map TK>

# Illustrations

# Introduction

## Himalaya, India, January 2008

At –16° C, it was warm for a winter's day in the high Himalaya. I was in the Spiti Valley of Himachal Pradesh, a huge gorge that cuts deep into the mountains of the greater Himalaya and Tibet. Wind-lashed, with temperatures dropping as low as –35° C, this is a high-altitude land of vertical rock faces and torrential glacier-melt rivers.

I had come to study the winter foraging behaviour of the blue sheep (*Pseudois nayaur*), a unique species of wild goat with sheep-like traits. Standing some 90 centimetres tall at the shoulder, the blue sheep live in herds which range as high as 19,000 feet above sea level. Superb climbers, they have an uncanny ability to 'disappear' into the slopes and cliffs, where their grey coats camouflage them against the rocky backdrop. When startled, they give a loud football referee-type whistle to alert their herd to danger. The males bear thick, ridged horns which curl sideways and back, whereas the females have thin, short, straight horns. Found across the Tibetan plateau and the greater Himalaya, these blue sheep are known by many names – *bharal* in Hindi, *yanyang* in Mandarin, *naur* in Nepali, *na* or *sna* in Tibetan, and more – and are the snow leopard's favoured prey.

My plan was to spend the winter of 2007 in Spiti tracking the blue sheep herds as they foraged, meticulously recording their behaviour and the plants they chose to eat.

That particular day I had left camp at six in the morning, and by half past seven reached the spot where I had seen blue sheep the day before – but there was no sign of them. I thought they might have gone beyond the next roll of the hill, but no such luck. After another two hours looking for my study herd, tired and cold, I decided to rest a while before resuming the search. I was now at the edge of a rolling plateau named Hom. On my right was a gorge, a vertiginous thousand-metre drop to the Shilla River, an important tributary of the Spiti. When a golden eagle flew below me in the gorge, I pulled out my camera to photograph this king of the Himalayan birds but it refused to start: its battery had been drained by the cold.

An hour of hard climbing through knee-deep snow took me to the crest of the plateau at an altitude of 4,500 metres. I gasped for breath in the rarefied air. The endless Tibetan steppe extended in front of me, the dark and cold gorge to the Shilla River on my right. Resting my weight on my ice axe, I was admiring the panoramic view when suddenly the cliffs below me came alive as a house-sized boulder fell with a crash into the gorge. By now I was used to seeing rock falls but never anything of this scale. A deep silence followed.

A chirpy whistle broke the quiet: the blue sheep alarm call. It rang from somewhere very close, but I couldn't see the animal. Then I spotted them nearby in the gorge – a group of twelve blue sheep grazing on one of the rock ledges only a hundred or so feet diagonally below me. My systematic observation of their behaviour and feeding could finally begin. Over an hour passed without much activity from the herd. The sky turned grey and feathery flakes of powdered snow filled the air and my body stiffened. The whole canyon was frozen – boulders covered with snow, a few leafless trees, no water flowing in the gorge – and the only sounds those of the occasional fall of smaller rocks into the abyss.

A small movement caught the corner of my eye. This was not a falling rock; there was no sound. The movement was on the same slope as my blue sheep and me. I looked carefully where I thought the ripple had come from. It moved again and this time I saw it: the ghost of the mountains. Snow leopard. He was so well camouflaged that I would never have noticed had he not taken a step forwards. The blue sheep were unaware of the snow leopard; the snow leopard was unaware of the blue sheep. He was on a ledge about 400 feet below me and his prey.

He must have been there the whole time. He seemed young, a little small to be a full-grown adult. He sat down briefly but stayed alert with his neck stretched, head up and eyes scanning the horizon. I was trying to pick out any peculiarities in his fur to identify him, but it was impossible to fix upon details at this distance. He got up and started walking along the cliff, moving gently, hardly leaving a footprint in the deepening snow. By now I was so familiar with these mountains that I knew exactly where that ledge would lead him. He would pass directly below the blue sheep herd I was watching, then below me, and come out on a wider slope further to my right.

Now the quest was to reach the starting point of the ledge before the snow leopard, and to hide there to get a better look. I ran at full tilt, struggling in the thin air. The shoulder of the plateau gave me complete cover from the snow leopard beneath. The start of the ledge on which the snow leopard was walking was slightly broad and had a little grass on it, so it was no surprise to find a herd of blue sheep grazing there as well. Seeing these blue sheep in the path of the approaching snow leopard, I was certain about two things: the snow leopard had not yet arrived, but when he did there was going to be a scene.

I took cover between two boulders, using one in front to prop my elbows like a tripod for my binoculars, and another behind to support my back. Hours seemed to pass, but still there was no sign of the snow leopard. The weather had turned windy and with the sun behind the clouds, the deep cold began to freeze the sweat on my clothes. I was considering whether the leopard might have turned and gone when, suddenly, another alarm whistle rang out: the snow leopard had arrived. He must be close – the blue sheep were running – yet I could not see him.

I had to think like a snow leopard: where would I be if I were him? I scanned the ledge along which I was expecting the snow leopard to arrive and eventually found him sitting under an overhanging rock, perfectly camouflaged. He was indeed a little young.

On many occasions I had seen two sets of pug marks in this area – the footprints of a mother and her cub. A camera-trap picture had shown the cub to be a male. Could this be him? Had he been weaned from his mother already? And where was she? Dead – or perhaps trying to mate again this winter? I was smiling with the sheer joy of seeing this incredible creature, but my mind was a roller coaster. Would he know how to hunt? Would he be able to make a kill any time soon?

And then, without warning, the snow leopard got up and gently glided down the steep sides of the gorge, disappearing into the boulders at the bottom.

Only then did I become conscious of my surroundings. An inch of snow had accumulated over the last few hours, a remarkable fall for one day in the Trans-Himalaya. The temperature had dropped further and only when I tried to move did I realise that my feet had frozen. I had been sitting still for too long. I found an overhanging rock, just like the one the snow leopard had used for shelter, gathered some twigs and lit a small fire. Taking off my boots, I gasped as warmed blood rushed into my feet causing intense pain – my toes thawing.

I sat for a while, my body aching with fatigue, but my heart full of happiness. Under that rock with the small fire to keep me warm, with the snow falling around me, I was overcome with emotions.

I had seen snow leopards once before, but this was special. For the first time, I experienced the perfect harmony between the mountains, the snow, the leopard, the blue sheep and me. I had spotted one of the rarest animals in the world in its natural habitat and accurately predicted its behaviour. The tranquillity of the experience told me I was not an outsider any more. I was one with the landscape and I felt accepted. It was impossible not to romanticize this encounter as an invitation into the world of the snow leopard. Studying the snow leopard had been a dream and now I had been initiated. I was alone with the snow leopard and the blue sheep, deep in the Himalaya, and in that moment I understood what it meant to be a snow leopard biologist and I knew what I wanted to do for the rest of my life. It is such moments of oneness with the mountains that I have kept seeking throughout my career.

This book is about a twenty-year journey studying the snow leopard. Since I began focusing on this rare animal, the most common question I have had from people around me, whether at dinner parties or meetings of policy directives or government bodies, is about the number of snow leopards alive today. It's an important question, since knowing the number of snow leopards might help unravel the causes of their decline and possibly even save them from extinction. A number can be fundamental to the conservation of a species and often draws more attention than highlighting the threats posed to its survival. A number gives governments a sense of control, a sense of understanding and a sense of authority. Counting the snow leopard has been a critical part of my scientific research, and what it takes to answer that question is a central part of the exploration of this book.

Yet the question about the number of snow leopards is nearly always followed by a question about whether I've ever caught sight of this most elusive of animals. Why do we care so much about seeing a wild snow leopard, and what fuels our desire to witness them in their natural habitat? Snow leopards have a special grip on the human imagination. In the early 1970s George Schaller, a wildlife scientist, became the first person ever to photograph a snow leopard in the wild. The picture was published with fanfare by *National Geographic*, just a few years after the Moon landing, and the snow leopard became a symbol of the mysteries that still exist on planet Earth.

Then in 1978 Schaller's friend and travel companion Peter Matthiessen published *The Snow Leopard*. In it, Matthiessen wrote of an expedition he had made with Schaller, shortly after his wife's death, to see the snow leopard in the Dolpa region of Nepal. Matthiessen never did see the snow leopard, but the creature's allure – its aesthetic beauty, its idyllic but desolate mountain habitats, its wary, almost mythical behaviour – draws us to this day. Shaped to perfection by evolution, snow leopards live solitary lives in the highest reaches of the Himalaya, surrounded only by glaciers, snow, rocks and icy peaks. They are shy, and their silvery-brown and white coats, marked with dark rosettes, work like magic cloaks offering such good camouflage that they can disappear in plain sight. They have a long furry tail – the longest of any cat in the world – which they wrap around their bodies like mufflers on cold nights. The tail also works like the balancing pole of a tightrope walker as they manoeuvre around exposed cliffs. Their thick, wide paws, lined with fur, act like snowshoes, helping them walk on snow like an elf. Their unusually large nostril cavities let them warm the cold air they breathe. They have small, rounded ears to minimize heat loss. They are silent predators, unable to roar. They move through the cliffs and rock faces like smoke. For me, nearly two decades and several sightings later, every snow leopard encounter is still a transcendent experience, like the crack of dawn after a dark night. Seeing the snow leopard changed the course of my life, and every new snow leopard sighting reminds me to go in search of the unknown and instils me with a sense of meaning and purpose.

The snow leopard is an endangered species, sparsely distributed in mountainous terrain in the twelve countries that straddle the Himalaya and Central Asia. They are legally protected in all these countries, but given their remote mountain habitat and inaccessible high elevation dwelling, their conservation did not gain urgency until recently. How could anyone prioritize the conservation of a species about which we knew so little? Knowing the animal would come before worrying about its survival. It was thought that they were safe in their rugged and isolated mountain abode, that humanity would never breach their Himalayan kingdom.

But climate change now threatens the glaciers and rivers of the Himalaya. Globalization has connected distant places into a complex web of supply chains across the planet, where consumer decisions in distant countries threaten the survival of species they have never seen and, sometimes, have never known existed. Extractive mining has reached the highest mountain peaks and the deepest oceans, and there is no protection even in the most remote corners of the Earth.

Yet the story of the snow leopard is one of hope. In my first encounters with the snow leopard, I mythologized them as part of a faraway world, distant from humans. But I soon came to realize how closely their lives are entwined with our human world. As my research progressed, I came to understand that I would be blessed with more glimpses of the snow leopard if I learnt more about the beings that surround them: the mountains where they live, the herbivores they hunt and the people who share this landscape with them. During nearly twenty years of research I have attempted to understand these entanglements as much as the biology of the snow leopard. The local people who are the real custodians of the mountains and its wildlife, and my time spent with them, make me a firm believer in a bright future for the snow leopard.

## Note on language

Throughout the book, I use the name 'Himalaya' instead of the more popular 'Himalayas'. Himalaya is already a plural of the Nepali word *himal*, meaning 'mountain'.

I have mentioned real people, and when I use a first name together with a second name, those are their real names. If I use only the first name that means I am using a pseudonym to protect their identity.

# PART I

# Mystery

# I

# First Encounter

'You will never see a snow leopard,' said a professor when I sought advice on my master's research project. 'And you won't get enough data to write your thesis.'

As a Wildlife Biology and Conservation student at the National Centre for Biological Sciences in Bangalore, the thesis I was due to present in 2008 was to be my gateway to a career in wildlife research. Modelled on American Ivy League universities, with acres of green lawns, air-conditioned research laboratories and cafeterias that serve cappuccino, the NCBS is one of India's leading institutes in biological systems research, studying everything from molecules to entire ecosystems. I had found my way into this elite scientific institution through hard work and a talent for wandering the wilderness, but now I was struggling to journey through the door I wanted to open – the one labelled 'snow leopard'.

'Your idea of studying predator–prey relationships between snow leopards and blue sheep in the Himalaya is very good,' said another faculty member, 'but you should do this research with the tiger and the deer instead, and work here in South India. Besides, it will be 40 degrees below zero during the winter and you are from one of the hottest parts of India, so how will you handle the Himalayan winter?'

Some of this was true. Only a handful of researchers – mainly from Western countries – had ever studied snow leopards in the wild, and they often struggled to back up their research claims with adequate data from the field.

It was also true that I had grown up in the heat, in a village in the central Indian state of Maharashtra, near the famous UNESCO heritage site of the twenty-nine Ajanta Buddhist caves cut into cliffs between the second and fourth centuries BC. My village often gets to over 40° C, and the hot summers of my childhood were spent eating mangoes and swimming in wells. I had grown up listening to my grandmother tell the story of the Last Tiger of Ajanta. This tiger had killed a cow on the outskirts of my village, and my grandmother recounted in great detail how my grandfather stayed up in a cowshed near the carcass and shot the tiger on a full-moon night. My father was a little boy then and remembers sitting next to the dead tiger, measuring his height against the length of the ferocious animal. My grandmother would end the story on a sad note, saying that was the last time anyone saw a tiger near our village. I grew up regretful of the outcome but tried to maintain a proud feeling that my grandfather had helped the village. The suggestion that I should study the tiger had resonance.

But I felt a calling to the Himalaya. I had been obsessively drawn to vertical surfaces since I was 15. Be they boulders, hillsides or buildings, I had to find a route to climb them. As a teenager I spent every waking moment thinking about how I could find my way to the Himalaya and be among the 'real mountains'. Living in the mountains, hiking without a destination and climbing was the dream – not to set records but simply to be with myself exploring my great passion. Doing my master's thesis in the Himalaya could locate me in the mountains for good. I knew I could handle the Himalayan winters. At 19, I trained at the Nehru Institute of Mountaineering in the Garhwal Himalaya of Uttarkashi, one of the best mountaineering schools in the country, and I could hold my own on icy slopes. During the NCBS's summer break, after the second semester of my current course, I had also interned with an organization working on snow leopard research and conservation, and I was confident I would gather enough data to complete my thesis. What I lacked, given an upbringing which emphasized an unquestioning respect for my elders, was the courage to say this to my professors.

The final decision would be made at the proposal defence seminar, a gruelling annual ritual at the end of the third semester at which students presented their research plans to an audience composed of professors and practising conservationists. If a student's idea was accepted, they would get the funding and permission to research at the proposed field site for six months. I had decided to propose researching the snow leopard's prey species, rather than the cat itself, in the hope that this would sound more achievable. I included a photograph of myself climbing an icy Himalayan peak to convince them that I could handle the winter cold. When I walked up to the lectern and the lights in the seminar hall dimmed, with faculty members and students leaning back in their seats to listen to my proposal, the voice ringing in my ears was that of a friend. She had told me that some people in the audience would support my plan simply for its audaciousness.

Two years before I started my MSc programme, I had made my first trip to the Himalaya. It was in 2004, and I had been accepted on the Basic Course at the Nehru Institute of Mountaineering in the Indian state of Uttarakhand. Named after the first prime minister of India, Jawaharlal Nehru, an avid hiker and trekker, the institute is India's premier mountaineering school. The principal of the school is appointed by the Indian Army from its ranks of senior officers who have led and excelled at climbing expeditions on the toughest mountains around the world.

After twenty-four hours on a train followed by an overnight bus, I reached the feet of the Himalaya at Rishikesh on a clear morning. Here, the great Himalaya rise abruptly out of the fertile plains of northern India. This is where the Ganga River comes rushing down from narrow valleys and spreads out over the wide, open space, calming down before meandering over the rest of her 1,500 miles to the sea. Uttarkashi and the institute were still a day's drive ahead but I was keen to experience Rishikesh before heading deeper into the mountains. This is where the Beatles had arrived thirty-five years before to seek spiritual enlightenment. Here John Lennon wrote the song 'Dear Prudence', inspiring people to reach out to someone who has isolated themselves, motivating empathy and human connection. For centuries Hindu pilgrims hoping to reach the Char Dham, the four sacred sites, have started their journey on the banks of the Ganga in Rishikesh before proceeding to the source of the Yamuna at Yamunotri and then to the source of the Ganga at Gangotri. The dedicated few hike the extra 18 kilometres to the snout of the Gangotri Glacier called *Gaumukh* or the Cow's Mouth – a place where the icy glacier melts into a river and becomes the River Bhagirathi, which later merges with Alaknanda to become the Ganga. It was these pilgrims who first noticed the retreat of the Gangotri glacier; for them, it was the literal reclusion of their spiritual site.

It was the high peaks beyond the glaciers that drew me in. I had read that I might be able to catch sight of the summits of Banderpunch (6,316 m) and Swargarohini (6,252 m) from Rishikesh. These stunningly beautiful mountains tower well over 20,000 feet and hold mythological significance in the Ramayana and the Mahabharata, the most revered Indian epics and foundational texts in Indian literature. Banderpunch, or Monkey's Tail, is named after Hanuman, the monkey god, and the story of how he leapt from India to Sri Lanka to save Princess Sita. But it was Swargarohini that held me in its sway. The name literally means a path to heaven. This is the route that the Pandava brothers and their wife, Draupadi, the protagonists of the Mahabharata, took to try and climb to heaven. For my 19-year-old self hoping to find a career in the outdoors, this would be my heaven on earth.

Even as I was getting down from the bus, my eyes were on the sky, looking out for the big mountain, and I stumbled a little. The chaiwala at the bus stop had seen this many times before. 'The mountains look nice from here, but it is very cold when you go up close. You should seek their blessings before you try to conquer them,' he said, before offering me a steaming-hot cup of tea. With a saffron scarf over his shoulders, a flowing beard and a calm face, he had the look of a wise man.

After two days in Rishikesh, I took a large shared cab to Uttarkashi, passengers paying by the seat. The taxi plied fixed routes, and the first person to pay took the front seat next to the driver. The next four got the middle and the next four sat at the back. The chaiwala had told me to wait to take a cab where I could get the front or middle seat. Within minutes of the car leaving Rishikesh, the road became windy, with cliffs on one side and a deep gorge on the other. As we climbed into the Himalaya, the River Ganga was a tiny white thread at the bottom of the steep green mountain walls. I was looking out for climbing routes on each boulder and rock that we passed. Every time the car crested a ridge and moved from one valley to the next, I saw a glimpse of the icy peaks in the distance. I could identify most of them by their shape, with the satisfaction of a birdwatcher identifying a bird they had never seen before. This helped me take my mind off those in the back seat who were carsick and throwing up out of the window. Every car we passed had rear windows laced with vomit. We had to survive eight more hours of this.

Our destination, Uttarkashi, was nestled deep in the middle Himalaya. Unlike the denuded slopes along the drive, the mountainsides here were covered in conifer trees, but they could not hide the three-pronged scar of the massive landslide that had buried a part of the town just the year before. Luckily, the residents had been evacuated and there were no human fatalities, but it was a reminder to everyone, pilgrim or mountaineer, of the dangers of venturing into the high Himalaya.

The morning after we reported at the Nehru Institute of Mountaineering, the students were split into small groups of five, called ropes, and assigned an instructor. I was part of rope number three and Naib Subedar Rajender Singh was our instructor. Naib Subedar was his rank in the Indian Army, but he had spent most of his career as a mountaineer. He was one of the summiteers from the recent Indian expedition to Mount Everest and Lhotse and he had traversed the ridge connecting the two peaks. He had been selected for the first Indian expedition to Shishapangma, the fourteenth highest peak in the world. I was being trained by one of the best in the business.

Rajender, sir, as we called him, reminded me of Sir Edmund

Hillary in his picture with Tenzing Norgay taken immediately after their return from the summit of Mount Everest in 1953. Rajender was tall with a long but pleasant face and weathered skin. He wore baggy clothes and a scarf on his head with the loose end around his neck. He spoke little, but every word he uttered was a piece of advice that could save my life while climbing. His first address to our rope began: 'You will work together and eat together and learn to trust each other with your lives. On the more dangerous sections of the climbing route, you will rope-up, which means be tied to each other with a rope and be responsible for your rope's safety. If one of you slips, the others have to anchor him down.' He always spoke in a matter-of-fact way.

Along with rope work, we would learn the three necessary skill-sets of mountaineering: rock craft, snow craft and ice craft. For the first eight days of the training, we would carry a fully loaded rucksack from the institute's campus to the Tekhla rock climbing area 8 kilometres away before spending the rest of the day climbing boulders the size of houses. Tekhla was a beautiful forest with well-weathered and rounded boulders strewn all over it. A landslide like the one in Uttarkashi town brought these boulders here hundreds if not thousands of years ago.

I had grown up wandering the black basalt hills around Ajanta, my village, so I was comfortable climbing boulders at Tekhla. I used this time to get to know my rope-mates and became good friends with Rajesh. He worked as a ranger with the Uttarakhand Forest Department. He was short, reaching only my shoulder, and not particularly stocky, and yet he was exceptionally strong and could easily anchor two of us pretending to slide off a steep rock face. He did not speak much during the first few days. Even later, he spoke only when there weren't many people around. He had joined the Forest Department at a young age and had spent a lot of time fighting forest fires. His job paid a pittance, and he was hoping to get a better-paying role within the same department by training as a mountaineer.

After finishing rock craft, we took a three-day hike with our rope-mates to reach the training base camp near the snout of the

Dokriani Glacier at an altitude of 14,000 feet. We walked through large stretches of conifer trees with patches of thick broadleaf forest. The sun barely reached the floor of these places. On the second day, we emerged onto slopes with strips of meadows. Large stretches of rhododendrons thinned out near the high ridges and icy peaks.

We were resting next to the hiking path in the middle of the day when I noticed Rajesh looking intently up a mountain slope. He did not shield his eyes with his hands or use binoculars; he held his head steady and stared with piercing eyes. I looked for whatever he was watching. At first, I could not see anything, and then, as if each layer were separating out, I could see the dirt, the rocks, the vegetation and the grey dots that materialized into mountain goats. They seemed pasted to the sides of the near-vertical rocks. Rajesh spoke in a soft voice. 'The *tangrol* can reach places which even mountaineers struggle to get to.' I could not pick out details at this distance. I was doing my undergraduate studies in zoology, botany and computer science at that time, and yet the word *tangrol* was unfamiliar to me. I asked Rajesh if it was the blue sheep or the ibex that he was talking about. He told me that they don't have ibex in this part of the Himalaya and left me to figure out the rest.

As we watched, more people joined us, but hardly anyone could spot the blue sheep, and they soon lost interest. I tried to guide them to the exact spot but to no avail. Soon, it was just the two of us again. I turned to Rajesh to ask something and was surprised to find the principal of the school beside me, Colonel Ashok Abbey. Rajesh was nowhere to be seen. At that time, Colonel Abbey was the biggest name in Indian mountaineering. He was the leader of the recent Everest–Lhotse expedition and was going to lead the first Indian expedition to Shishapangma. He had a stern face, and I was expecting a reprimand for being distracted by the blue sheep, but to my surprise, he said, 'Good spotting.' Before I could tell him that it was Rajesh who had spotted them, he added that the school was carrying a library of books weighing 60 kilos to the base camp. I should try to find a book about the blue sheep. By now, I was familiar with the mountaineering way of measuring all things by their weight. Later, Rajender told us that spotting blue sheep is helpful because they can set loose rocks rolling down on the climbers below. Blue sheep like to stay on higher ground than people because it gives them a sense of security. Little did I know that in a few years' time, the blue sheep would be at the centre of my research. What's more, I did not know that where there are blue sheep, there are snow leopards.

Our base camp was on a flat area away from the risk of avalanches and resembled a small village. Forty students were divided among four large tents with two ropes in each tent. The instructors were housed in five small two-people tents. There was one large mess tent and one ration tent where the library was also housed. The principal had a tent of his own a little further away. We could see the snout of the Dokriani Glacier from our tents. The river originating from this glacier gushed past our camp at a distance of about a hundred metres. Across the river was a peak called Draupadi-Ka-Danda (5,670 m), meaning Draupadi's walking staff. The glacier originated at the base of Banderpunch, the peak that I had seen from Rishikesh.

It was October, and there were a few inches of snow at the base camp. For the next two weeks, we would train in snow craft on the slopes of Draupadi-Ka-Danda and ice craft on the Dokriani Glacier. Every day we would head out to the mountain after an early breakfast and train until noon, when the rising sun would melt snow and ice, triggering rock falls and avalanches. Theory classes filled our afternoon and evenings, where we learnt first aid, expedition planning, high-elevation acclimatization and navigation. This training would culminate in a summit push during the last two days to Draupadi-Ka-Danda or one of the other high points in the surrounding area. Rajender had built up the summit push as the test that would separate the hikers from the climbers, but he was measured in saying this was an expedition and not a competition and he expected the rope to function as a single unit. The summit push was a rite of passage for every batch of students, but eighteen years later, on 4 October 2022, a batch failed. An avalanche high up on Draupadi-Ka-Danda claimed twenty-eight students and two instructors. It was arguably one of the worst mountaineering disasters anywhere in the world but this news barely reached the global media. The world was still recovering from the pandemic, and nobody wanted to hear about more deaths in India.

I had taken our principal's advice seriously and made a beeline for the ration tent on the first day of settling into the base camp. The 60-kilo library was lined up on makeshift shelves made of wooden slats surrounded by cartons of sugar, rice, canned meat and flour. The main flap was the only source of light and air in the large tent made of military green fabric. The library contained the who's-who of the mountaineering world. Lining the shelves were accounts of Sir Chris Bonington's team climbing the south face of Annapurna, Maurice Herzog's first summit of an 8,000-metre mountain, Hermann Buhl's first ascent of Nanga Parbat, Captain M. S. Kohli and India's first successful expedition to Everest. I could not find anything about the blue sheep but something else caught my eye: *My Quest for the Yeti: Confronting the Himalayas' Deepest Mystery* by the legendary Reinhold Messner. Messner is one of the greatest climbers of all time. He was the first person to climb all fourteen peaks higher than 8,000 metres, and he did so without supplementary oxygen; he had walked across Antarctica and Greenland, and made a solo crossing of the Mongolian Gobi Desert, and the book's title suggested that he was obsessed with the mythical yeti. At first, I was surprised that the greatest mountaineer of all time went about searching for the yeti, which I, as a zoology student, understood to be a fabrication of the human imagination. But the suggestion of bigger unknowns than unclimbed peaks and untried routes caught my fancy. I checked out the book by informing the mess manager. I made a mental note to keep it hidden from the principal and Rajender because I did not want them to think that I was not serious about mountaineering. A picture of Reinhold Messner hung at the institute's main building, and I had heard on the grapevine that the principal knew him personally. It crossed my mind that I could ask the principal about Messner and his thoughts about the yeti but the risk of being dismissed was too great. I had to do my own reading first.

Ice craft started with us learning to walk with crampons on the surface of the glacier. Having spikes on my feet was harder than I imagined, and also dangerous if I did not master it soon enough. Once we became used to them, we started climbing ice walls with an ice axe in one hand and an ice hammer in the other. The list of things to keep in mind kept growing: an axe and a hammer in both hands, pointy crampon toes, a belay rope that goes through different pitons anchored to the ice wall, colleagues climbing above you who might accidentally dislodge loose rock and ice, colleagues climbing behind you whom you might end up accidentally hurting, spare a thought for the ever-changing weather – and the list went on. But, gradually, much of this became muscle memory.

I was training during the day and reading about Messner and the yeti in my sleeping bag at night. My progress on the ice was faster than my progress with the book. The workload was tiring and I needed enough food and rest to keep up. In the early days, many students had reduced appetites because of the altitude. Rajender would say at meal times, 'The army marches on its belly and mountaineers crawl on theirs.' But each page of Messner's journey to unravel the mystery of the yeti was gripping. Messner was exploring the route that the Sherpa people believed their ancestors took to get from Eastern Tibet to the Khumbu Valley. One evening, in fading light, he stumbled upon a creature larger than a human and walking on two legs. He then saw its footprints, which were bigger than anything he had seen before. He wrote about a photo of a supposed yeti footprint taken by the pioneer Himalayan mountaineer Eric Shipton in 1951, which made the yeti popular in the media of the time. The rage was such that even Sir Edmund Hillary led an expedition in search of the yeti in 1960.

'To avoid being driven insane by this mystery [of the yeti], I would have to return to the mountains', Messner wrote.

I was mesmerized not by the yeti but by how some of the leading mountaineers of the world were utterly captivated by the story of the mythological creature. I asked Rajesh about the yeti, and he dismissed the idea without a second thought. He then went on to describe the behaviour of the two species of bears that are found in the Himalaya. The Himalayan black bear is found in the lower and middle Himalaya in the forested habitats, while the brown bear of the alpine meadows lives closer to the snow line. I asked him which one of the two would be more dangerous and he said, 'They are the same; they will run away if you warn them of your approach. But more people are injured by the black bear because they stumble into them in the closed forests. The open vegetation of the higher reaches gives people and the brown bears enough time to see and avoid each other.' Rajesh spoke with such certainty and experience about Himalayan animals that I knew never to bring up the yeti again.

The base camp was on edge with nervousness the night before the summit push. At evening teatime, some people were unusually quiet while others spoke too much. The principal addressed the camp and announced that we would be attempting an unnamed peak of about 19,000 feet. Draupadi-Ka-Danda and Banderpunch were considered too dangerous for a batch of first-timers due to the heavy snow. Rajender gave us a pep talk. 'Tomorrow, you will turn a new page in your lives,' he said. 'Either you will give up climbing for good, or the mountains will become a part of your lives forever.' We listened to him with rapt attention; his words were divine knowledge. He warned us that most accidents happen on the way down when our minds start to wander with pride at having claimed the summit or from the disappointment of failure. 'Stay together as a rope and don't drop your guard until everyone is back in the camp,' he added. He wished us success in our careers as climbers and, as a parting thought, said that we should always know when to turn back. 'No mountain summit is worth your life; don't make last-ditch attempts. Come back another day and try again.'

Later in the evening, it was dusk in the valley, but the summit of Banderpooch was alight with the last rays of the setting sun reflecting off its icy face. I left for a short walk to help calm my nerves. My mind was alive with thoughts about mountaineering and the yeti. A little way outside the camp, I felt as if I were being watched, so I turned around. A bulb was glowing near the mess tent, but otherwise everything was quiet. The thought of the yeti briefly crossed my mind. I walked a little more before sitting atop a boulder, watching the sunlit summit of Banderpunch slowly sink into the evening. The summit turned from fiery orange to icy blue. I thought about all the mountaineers who had ventured into places like these and those who had perished, sometimes without a trace. I remained sitting until my mind was at peace. It was time to get back to camp. There was still ambient twilight but I felt for my headlamp in my pocket in case I needed it.

Something felt off on the trail ahead. I noticed a pug mark on the trail, a spoor on top of my shoe print, crossing the path perpendicularly from the stream on my right to the high ridge on my left. A thick oblong pad, the size of the base of my palm, and four perfectly round toes without claw marks. Dog pug marks always show the marks of their claws because they cannot retract them like cats. These marks had to belong to a cat, and this high up in the Himalaya it could only be a snow leopard. I switched on my headlamp and looked around, imagining a snow leopard staring back at me with curiosity. I was in the same spot in which I had had the feeling of being watched. Was the snow leopard hiding by the trail, waiting for me to move on before he crossed it and went up the ridge?

I returned to the camp high on the excitement of having seen the pug mark. I could now relate to Messner, who believed he had seen a yeti; out there in the remote mountain landscape, the presence of wild animals is keenly sensed. Rajesh was in his sleeping bag. I did not want to disturb him but he peeked out and asked what I was so excited about. I hesitated before telling him about the pug mark. He wanted to see it. So we sneaked out of our tent and went down the trail without our headlamps to avoid attracting attention, especially from the instructors and, most of all, the principal.

I had a great sense of validation when Rajesh confirmed that it was a snow leopard pug mark. 'If there is still a mystery left in the Himalaya, then it is the snow leopard,' he said. 'I believe I saw one a couple of years ago when we were hiking in the Gangotri region but I never told anyone about it out of fear of being ridiculed. I could have gotten away with saying I saw a yeti, but claiming to have seen a snow leopard would have made me the butt of all jokes.' We spoke about snow leopards late into the night. Rajesh had grown up in the Himalaya and worked for the Forest Department yet he did not know another living person who had seen a snow leopard. Their shy behaviour and camouflage makes them particularly hard to see. Their ochre-and-dust-colour coat with its irregular rosettes looks exactly like a rock covered with lichens. Their small, rounded ears create a soft and indistinct shape to their head, and the rosette fur breaks their outline. The only way to see a snow leopard is to spot its silhouette against the snow or its movement against a still mountainside. They struggle to hide themselves on snow. They are most comfortable on rocky crags, disappearing at will and climbing like apparitions. They should really be called rock leopards not snow leopards, I learnt from Rajesh.

Rope three reached the highest point, somewhere above 18,000 feet, before we turned back due to inclement weather. Nobody made the summit that day. Rajender was proud of his rope's performance and told us that we had surpassed the principal's expectations. And when I asked him about his time on the Everest expedition, he said, 'Everest is for tourists. There is nothing new to be done on Everest. Go find a mountain that calls your inner mountaineer. A real achievement needs a meditative understanding of oneself and the mountain.'

I was not yet ready to scale a vertical rock face on an 8,000-metre mountain by myself, but I was sure that someday I could lead a team of climbers on a significant mountaineering expedition. I was ambitious and wanted to accomplish great climbs, but Rajender helped me focus on skills, mastery and perfection. He had shown us that great feats of climbing are achieved not through recklessness and bravado but through careful planning, discipline and teamwork.

Rope three felt very good about ourselves after the summit push and we were already thinking about the summits we would climb in our mountaineering careers, but my mind was also rushing back to the pug mark. Along the climb, I had been looking around, scanning the slopes, hoping to glimpse the snow leopard – while knowing full well that a snow leopard would not come close to an expedition of more than forty climbers. I was haunted by the pug mark in the same way that Messner was with the yeti. I wanted to trek the Himalaya endlessly in search of the snow leopard. I never thought I would become a scientist, but I knew that chasing the unknown would lead me to a life of adventure.

I went back down the trail one last time to see the pug mark before we wound up the base camp, not knowing if I would see another in my lifetime.

# 2

# Sunshine and Shadow

My Basic Course in Mountaineering was subsidized by the government and barely cost anything other than the train and bus tickets. I immediately went on to do an Advanced Course in Mountaineering from the Himalayan Mountaineering Institute in Darjeeling, set up by Tenzing Norgay, Edmund Hillary's partner in the first summit of Everest. By this time I was a seasoned climber, and I was looking for a job in the mountaineering world. I had written to the Indian Mountaineering Foundation in Delhi for opportunities.

I lived in a small town called Aurangabad, close to the geographic centre of India. The town had seen glorious days in the past when the last of the important Mughal kings, Aurangzeb, moved his military headquarters there in 1681 to control the Maratha kings of this region. The city is built inside and around the stone fort of the ancient military headquarters. About three hundred years before Aurangzeb, the Delhi Sultan Muhammad bin Tughluq had moved the capital of his empire for a brief period of seven years from Delhi to the Devgiri fort, just 20 kilometres from Aurangabad. But by the early 2000s, it was a dusty little town with few job opportunities, nor did it have the charm of a historical city like Venice or Naples. Cycling around town, one passed through beautiful stone arches that were falling apart and historical buildings that lay abandoned or were being reconstructed with little thought to cultural preservation. Mountaineering was going to be my way out.

Climbing was not something that I did impulsively to right the wrongs of life, as there was nothing particularly wrong in my life. Climbing was a calculated idea to help me escape the monotonic numbness of small-town India. For most of my friends, their route out was to become an engineer or a doctor or to work for a well-paying corporate in a larger city. Such a career for me would have been jumping from the frying pan into the fire. If mountaineering failed me, I was prepared to go back to my roots and be a farmer in my village.

Aurangabad had a free public library tucked on a small street close to the centre of the town, right next to the butcher's lane. The stench of partially digested animal excreta mixed with blood filled the air. The library was a circular building, painted white, with a central round courtyard. One half-circle of the building was filled with bookshelves and the other half had desks for people to sit at and read. A newspaper stand was near the entrance. The library was usually frequented by students aspiring for the civil services. They would bring their own books to study but use the library for peace and quiet away from their homes or colleges and workplaces.

If I was not climbing some rocks in one of the many quarries that surround the town, I would usually be found in the library sifting through its catalogues, looking for books about anything and everything related to mountains. Here, I found the books that shaped my life. First was Jon Krakauer's *Into Thin Air*, about the 1996 disaster on Mount Everest, when eight climbers lost their lives. Krakauer put the blame on commercial expeditions in which rich clients paid professional mountaineers to help them reach the summits of the world's highest mountains. I read the response to this book by Anatoli Boukreev, one of the mountain guides who had saved three climbers but was blamed by Krakauer for having only helped clients from his own company and left some of the others, who later died. Anatoli Boukreev was one of the strongest climbers in the world at that time. From Soviet Kazakhstan, he had already summitted ten of the fourteen 8,000-metre peaks, but found him-self a victim of the economic conditions following the break-up of the Soviet Union, so was earning a living as a mountain guide. That fateful night on Everest, he helped three climbers down from the death zone wearing only a pair of hiking shoes. Both a hero and a villain in the disaster of Everest in 1996, he died in an avalanche on Annapurna the year after. *The Climb*, his response to Jon Krakauer, was published only a few months before his death.

I was thrilled by the adventures described in these books. I imagined myself heading up a mountain in my sneakers to save a fellow climber's life. In my youth and inexperience, the sudden accidents and almost freakish ends to the lives of great climbers did not bother me. I read some of these accounts so many times that I could recount the climb from the base camp to their summit, each step of the way.

Another book I read was George Schaller's *The Year of the Gorilla*. Schaller, a German-American scientist, was a young man in the 1950s studying gorillas in the Virunga mountains along the Zaire–Rwanda–Uganda border in Africa. He loved the solitude of the naturalist. His observations of gorillas showed the world that they were not the aggressive animals we imagined them to be; rather, they lived in tight-knit social groups forming caring societies. Schaller's description of the natural history of the Virungas reminded me of the snow leopard pug mark I had spotted at base camp near the Dokriani Glacier. Could I be somebody who studied the snow leopard in the Himalaya? It was a big dream for a young person sitting in a corner of a public library in Aurangabad.

Two years later, in 2006, I was selected for the master's course in Wildlife Biology and Conservation at the National Centre for Biological Sciences in Bangalore. The course admitted only fifteen students every two years, and selection was made through a written exam followed by an interview. Guest faculty from some of the best universities in the world came to teach. The course was divided into four semesters over two years. The first three were theory semesters and the fourth was a field-based thesis. I dreamed of being like Schaller studying gorillas in the Virunga and that the university would offer me an opportunity, if I could convince them of my plans. While most faculty members believed that studying the snow leopard through the six months of winter for a master's dissertation was too difficult due to the elusive nature of the animal, they were willing to hear me out if I talked about the blue sheep, one of its primary prey species found in the same habitat.

The blue sheep is the most common herbivore of the Tibetan plateau and the high Himalaya and lives above the treeline. It plays an important role in the regulation of these ecosystems, both as a herbivore that eats the vegetation and as a prey for carnivores like snow leopards. Their commonness meant that I could easily find them and collect enough observations to write a thesis, and my research findings would be relevant not only for the blue sheep but also for the ecosystem and the snow leopards. I would study the snow leopard by proxy.

The blue sheep is one of about thirty-five species of goats, sheep and their relatives from the sub-family Caprinae, which includes all the mountain-dwelling herbivores found around the world, including the musk ox of the Arctic, the chamois from the Alps and our humble domestic goats and sheep. The caprinae diverged from their relatives, the antelopes, 15 to 18 million years ago. The antelopes occupied the various different landscapes of the deserts, savannah and forests of Africa, but Africa does not have big alpine habitats. The summits of Mount Kilimanjaro, Mount Kenya and a small portion of the Atlas mountains in Morocco are covered in snow for a brief period annually. It was only when the ancestors of the goat-antelopes reached Eurasia that they encountered the big mountains and radiated into many species.

All along the mountain ranges stretching from western Europe through Asia into North America, different species of wild goats and sheep have adapted to living in a harsh vertiginous world and different predators have evolved ways to hunt them. Most often, the same carnivore that hunts in the plains also stalks the mountains but employs different hunting methods. The wolf and lynx in Europe hunt deer in the plains and ibex and chamois in the mountains; Persian leopards do the same in the Middle East; the puma in North America. The snow leopard, however, is the only species of large carnivore that specializes in a life in the mountains. It never makes a home in the plains. Snow leopards are limited to the highest mountains of Asia, from the Himalaya in the south through the Zanskar, Karakoram, Pamir and Altai in the north and along the Tibet–Qinghai plateau in the east. In these vertiginous lands, they specialize in hunting mountain goats and sheep, by either using their long tails to help them balance on steep rocky cliffs to catch the goats or ambushing the sheep in narrow gullies along the rolling hills.

A close examination of the snow leopard's anatomical structure tells an interesting story. Their forelegs have great flexibility to hunt in steep rocky terrain. Their eyes – bigger than leopards' – are placed more widely for better stereo vision to help with searching for prey in the open landscape. They have large cheek teeth that can slice through frozen meat. Their jaw and canines wield a slow but strong grip to hold on to the goats and sheep that they hunt on steep slopes. They have a wide nasal and sinus cavity to breathe in the thinner and colder air of the high elevations, especially necessary when hunting in tough terrain.

Fossil evidence suggests snow leopards may have occupied every mountain range from Portugal to Siberia during the last ice age. Paleontologists are still trying to understand the historical distribution of snow leopards and how the global climate affected them in the past and how it may affect them in the future. What is certain is that they were always at home in the mountains hunting goats and sheep.

Studying the blue sheep for my master's dissertation would not only bring me close to the snow leopards but would also open a door onto the evolutionary past of the goats and sheep that have come to colonize almost every mountain range in the world. There had been a few studies on the ecology and behaviour of the blue sheep and I would have enough background information to base my research on.

I planned to specifically study the feeding behaviour of the blue sheep, because that would tell me how they survive the long harsh winter in the Himalaya when the plants have withered away or are buried under the snow. From a herbivore's point of view, plants of the high Himalaya can be classified into two main groups: grasses, and herbaceous flowering plants also known as browse. Grasses and herbs deploy different strategies to deter a herbivore like the blue sheep from eating them. Grasses are coarse due to their high silica, which makes them less nutritious. Herbs deploy secondary compounds that can taste bitter and are sometimes poisonous. Herbivores have specific adaptations to digest grasses or herbaceous plants. Switching between the two can be difficult even for the most hardened species.

All the literature that I had read about the blue sheep suggested that they are adapted to eating grass. They have a relatively wide mouth and prefer grassy slopes to feed on. Studies of their summer diet found that up to 80 per cent consists of grasses. However, one study of their winter diet said that they ate more browse during the winter. Our theories of diet in herbivores tell us that herbivores should not switch between grass and browse from summer to winter. So why were the blue sheep doing it? Why should a grazer browse?

This was my primary research question at the thesis defence seminar. The main hypothesis was that livestock such as domestic goats, sheep, cows, donkeys, horses and yak, and their hybrids with cows, many of whom are grazers, deplete grass availability during the winter, leaving the blue sheep with no choice but to feed on the leftover browse. This, I believed, could affect the population of the blue sheep. Would they be able to survive the winter and reproduce as effectively if they did not have access to their preferred food?

Thinking about the research problem was easier than designing a study to answer my question. My audacious plan was to simultaneously work at three different locations that had a gradient of intensity of livestock grazing: one site with high livestock grazing, which would have depleted grass availability in winter; one with moderate intensity of livestock grazing; and one without any livestock. My plan was to measure grass availability at each of these sites, observe blue sheep diets, measure the nutritional value of both grass and browse in each of these places and, finally, at the end of the winter, count the number of female blue sheep and young that were born at each of these sites. If my hypothesis was correct, then the blue sheep would eat grasses 80 per cent of the time in the site without any livestock, and at this location they would also have more young born per female. Knowing whether changes in blue sheep diets have population-wide consequences means we are better able to conserve the species.

There were many questions after my presentation but most were about the details of data collection. Nobody questioned the idea of a thesis in the remote Himalaya at master's level. Half the battle was won.

For the other half – locating the right place for my fieldwork – I was planning to stand on the shoulders of giants. Dr Charudutt Mishra and Dr Yash Veer Bhatnagar of the Nature Conservation Foundation (commonly known as NCF) were pioneers of studies of mountain goats and sheep in the Himalaya. Yash Veer-ji, as I called him, had done his PhD studying the Himalayan ibex, and Charu's PhD was on blue sheep and their interactions with livestock. I spent a month interning with them during my summer break at their field site in the Spiti Valley in the state of Himachal Pradesh. My thesis ideas were shaped by discussions with Charu and Yash Veer, and they helped me identify three locations in the Spiti Valley with a gradient of livestock grazing.

My fourteen classmates had equally grand plans of their own, and after the thesis defence seminar, we were ready to spread out across thirteen different states of India, studying wildly different species and ecosystems: from the striped hyena in the deserts of western India to rare birds in the rain forests of the north-east. We parted with promises of writing letters, as most of us would be doing our field research in places not yet connected with reliable phone networks.

The Spiti Valley is a wide region on the northern slopes of the Himalaya between the mountains and the Tibetan plateau. Historically, the traders who moved between the plains of north India and Tibet called this place the Middle Land. Geographers, however, call it the Trans-Himalaya. The Trans-Himalaya does not receive monsoon rains. Even the mighty monsoons cannot cross over the high Himalaya, leaving it a parched, cold desert.

The little moisture that it receives comes from the Mediterranean Sea, often in the form of powder snow in the winter. The pastures are habitats to a mix of grasses, flowering plants and a few stunted thorny shrubs, like the famous wild rose. My plan was to live in a tiny hamlet called Tashigang and track the blue sheep in the three pastures around it. It had five houses and a total of seventeen inhabitants.

Before spending the winter cut off from the rest of the world, I visited my parents and my sister in Aurangabad. From Aurangabad, the first leg of my journey to Tashigang was a thirty-six-hour train ride to Ambala, a long journey which was delayed by a further twelve hours along the way. Train rides in India are a social affair, and I spent most of my time answering fellow passengers curious to know why I was going to the Himalaya with a rucksack at the start of winter. At Ambala I changed to a bus that would take me to Shimla, a town known as the queen of the hills and British India's summer capital. After three days, I had barely made it to the foot-hills of the Himalaya. I stayed in Shimla for a day before the next stretch of the bus ride that would take me to Kaza, the administrative headquarters of Spiti. For this stretch of the journey I would be travelling across the greater Himalaya, but instead of crossing over a high pass, the road follows the Satluj River which cuts a deep furrow from its origin at Manasarowar on the Tibetan plateau south across the mountain range.

It was 6 p.m., and before boarding the bus, I climbed onto its roof to secure my rucksack with all my equipment and clothing necessary for six months of Himalayan winter. It was my responsibility to ensure that it did not fall off, get lost or get wet if it rained. I used every strap available on the rucksack and two additional bungee cords and hoped for the best.

I asked the ticket conductor how long it would take us to reach Kaza and if I could get a seat at the front of the bus because I get motion sickness. He smiled at me. 'Don't worry, almost everyone on this bus will get sick; the best thing will be to sleep as comfortably as you can. Feel free to use the aisle to lie down after we stop for dinner,' he said in a comforting voice. 'We will get there when we get there. It depends on the roads,' he added as an afterthought. 'I thought it depended on the driver,' I said to myself, but in the years to come, I learned that the roads have to be maintained every day to be operational. A few hours of rain or snow would trigger landslides that could take days to fix.

I slept through most of the bus ride as advised by the conductor. The few times I peeked out the window, all I could see were granite walls on both sides following a stringy muddy river. At times we were close to the river, and at others I was scared to look down, wondering how far we would fall before hitting the water. It took us twenty-four hours to cover the 400 kilometres to Kaza.

Kaza had more government offices than houses. The government buildings were dull grey cement blocks with sloping tin roofs. The local houses were beautiful mud buildings painted white with square windows on either side of the front door. All were two-storied with larger windows on the upper floor and smaller windows on the lower. The roofs were flat like the houses in my village, and plastered in mud and lined with shrubs to prevent rain and snow from etching the edges. The village sat on the banks of the Spiti River at an elevation of 3,500 metres. It was 3 December, and winter was already here. The streets were frozen and slippery and towards the south of the village, rock and scree slopes merged into snow and ice, eventually reaching ridges above 5,000 metres and connecting mountain peaks that were not visible but stood well above 6,000 metres. I peered up into the landscape, searching already for climbing routes to find the snow leopards.

After two nights in Kaza, I was acclimatized to the altitude and anxious to reach Tashigang. It was the seventh day since leaving home. Two people I knew from my internship came looking for me, Sushil Dorje and Tanzin Thinley. They were both tall, and Thinley, who was older than me, looked like he was just out of high school. Sushil Dorje had assisted Charu through his PhD and Tanzin Thinley had only recently joined NCF's team in Spiti. They had brought a car with them to drive me to Tashigang. We packed supplies before we headed out, bought from the only grocery store in Kaza. With little choice in the store, we bought a sack of tomatoes that we would freeze and use as needed, and sacks of potatoes, rice, lentils and wheat flour that we would have to keep warm and dry. Then Sushil turned to me and said casually, 'We will buy the yak in Tashigang.'

I thought I had misheard him, so I asked again.

'You will need protein and fat to survive the winter at 4,400 metres. Where else do you plan to get that?' I was reminded of Rajender's maxim that 'mountaineers crawl on their belly'.

During the three-hour drive from Kaza to Tashigang over steep uphill roads we made small talk and Sushil told me that his real name was Tandup Dorje. He was called Sushil because his teacher at school was from the plains of north India and could not pronounce his name correctly. Now everyone, even his wife and kids, called him Sushil.

The loaded car moved along slowly in its strongest gear. As we moved up from Kaza, the valley opened up. We passed Kibber, where Thinley and Sushil lived and where I had spent my internship days. They would return to Kibber before night fall after setting me up in Tashigang. Tashigang was perched on a plateau north of Kaza and all of its five houses faced south, overlooking a small frozen stream. I could see the village from a kilometre out, but as our car came around a bend closer to the village, the beautiful pyramid-shaped mountain Chau Chau Kang Nilda also came into view, standing well over 6,300 metres, the upper half made of ice, the lower half made of rock.

In Tashigang, I was to make camp in the spare room on the upper floor of one of the five houses. It was a small room, about 12 by 12 feet, painted in bottle green with a black metal wood-burning stove in the middle. A large window on the wall opposite the door opened onto a mountain view. We set three mattresses around the stove, with three low tables called *chokse* between the mattress and the stove, in the customary Spitian way. I spread out my prized sleeping bag on the bed directly behind the stove, claiming that spot. I had spent nearly half of my research budget on this imported sleeping bag. Six months would be a lot of cold to endure. The toilet was next door to the room. It was a Spitian dry toilet, essentially a hole in the ground. The waste accumulates in a dark room below, where it decomposes and becomes fertiliser in time for spring in June. The toilet only had a curtain. Sushil saw the look on my face and said, 'Don't worry, just ask before you enter. Everyone does that.'

Sushil and Thinley were planning to return to their village the same day. They had hired a young man, Kalzang Palzor or KP, from the neighbouring village to help me with my research at Tashigang and life at what we were already calling the 'base camp'. Sushil introduced KP as someone who would ensure I didn't die while chasing the blue sheep. KP occupied the mattress next to the window. The window had only a single pane of glass, so we fixed a thick sheet of plastic over it to keep the freezing wind from entering the room through the cracks in the wooden frame. I had lived in many different hostels throughout my school and college education, but this room became the first place outside of my parents' house that I would consider home. It wasn't much, but this was where I would return to after long days of fieldwork, lick my wounds after misadventures and read the letters I received from friends and family, over and over again, through the long and cold winter storms.

KP made chai to bid farewell to Sushil and Thinley. They promised to return in three days to check on me. Just then, the door opened without warning, and a tall man with a thin moustache poked his head in. 'I can smell some tea brewing,' he said. Sushil invited him inside and introduced him as Thukten Kalzang, the schoolteacher who taught the three kids in the village. 'He walks 7 kilometres each way from his village to Tashigang every day,' Sushil said proudly, which made my planned daily task sound simple in comparison. Sushil introduced me and my upcoming research project on the blue sheep.

The teacher was a cheerful person with a smiling face. He asked who was going to use the third mattress, and when we said there was no one else joining the team, he decided to move in since there were two new youngsters in the village, and he no longer needed to commute back and forth. Sushil, Thinley, and KP nodded as if this were the obvious thing to do. I followed their lead. I was willing to do whatever it took to be accepted in this little hamlet. If it meant sharing a tiny room and a toilet without a door with the local schoolteacher, then so be it.

The teacher proved to be an excellent cook; he also took charge of splitting wood for our base camp, which he was already doing for the school. He took an interest in my data sheets and promised to help make copies as backups. He had a radio that we would sometimes listen to. It became our main source of news from the outside world. One day we heard that schools in Delhi were shut because of an intense cold spell, but Tashigang's school functioned every day even when the temperature was 30 degrees below zero. The teacher was full of energy, and within days I was grateful that he had made himself at home in our base camp.

Thinley and Shushil returned three days later. I was not yet fully acclimatized and was breathless every time I worked hard. I had not seen a blue sheep, so they decided that I needed a short walk to introduce myself to my study subjects. They told me it would be good for me to stretch my legs and that it could help with my acclimatization.

We had barely walked half a kilometre when we heard a loud crack, not quite like a gunshot, more like the snapping of a thick wooden branch. I stood still mid-trail and Sushil ran past me, saying, 'The rutting is still on.' We rushed after him and over the next rise of the hill, where about sixty blue sheep were standing. In the background, the pyramid-shaped Chau Chau Kang Nilda glistened in the mid-morning sun. Two big males were ramming each other with their horns and the cracking echoed from the cliffs.

The females grazed without lifting their heads. The fifteen or so large males sized each other up, and every once in a while one of them, usually the larger sheep, would headbutt the smaller one – more often on the flank or the rump; only similar-sized ones fought head-on. Their grey coats were fresh and thick, ready for the winter. The males had darker hues and the young ones were lighter than the females. The largest male in the group was very busy. Between bouts of headbutting the other males, he would walk through the herd with his head held low, neck stretched out and, based on smell cues, would approach a female from behind and then lift his head high, curl his upper lip and make a grimacing face. I knew this to be what is called the Flehmen response. The male was channelling the pheromones of the female to a special chemical receptor called the Jacobson's organ located on the roof of his mouth to detect if the female was in oestrous. If he found a female in oestrous he would follow her around while keeping the other males back, and mate with her multiple times. This, I had read, was the most common mating tactic for males competing for access to females to pass their genes to the next generation. However, it's a tactic that only works for the largest males in the herd who have the physical strength to keep four, five or sometimes even six competing males at bay. The smaller males tried a different tactic called sneaky mating. As the name suggests, they try to sneak up to the female while the big male is busy warding off the competing males. In some species of fish, the smaller males change colour and shape to mimic females to sneak past the big males who challenge any males getting close to the females.

Evolutionary biologists are interested in understanding how these different mating tactics coexist within a population. It is an interesting question, but I was more curious about a proximate problem. The males seemed to be exhausting themselves with all this fighting, following, coursing and sneaking, and they had a five-month-long winter ahead when there would hardly be anything to eat. How were they going to survive the winter? I was also interested in the females because most of them were carrying or would get pregnant soon. The developing foetus would need a lot of energy. How would they carry their baby to term through the hard winter?

This was my first up-close encounter with the blue sheep since deciding to study them for my thesis. Having read all the previous research about them, I was able to notice many interesting behaviours within a matter of minutes. They were not merely sheep-like goats any more, they were my centre of attention – my window into understanding how animals survived the Himalayan winters. Earlier when Sushil pointed to them with an open palm and said, 'There are your blue sheep,' as if handing me a responsibility, the words 'your blue sheep' lingered in my ears a long time.

I knew I wanted to study the snow leopard some day, but I was in love with the blue sheep right now. My head was full of questions, and the thought that I would be studying these tenacious and resilient animals in this austere yet majestic place gave me goosebumps.

All this while, we were standing barely 70 metres from the blue sheep who were hardly bothered by our presence. Sushil told me that we had caught the tail end of the mating season, so the males were busy and distracted. The people of Tashigang don't harm or disturb the blue sheep. The blue sheep must think that we were from Tashigang. I liked the sound of that. As long as they thought I was from Tashigang, they wouldn't be scared of me.

On 3 December, ten days after leaving home in Aurangabad, I was in front of wild blue sheep, making notes on their behaviour like Schaller did with the gorillas fifty years ago. My library dream had come true.

In another week, I was well acclimatized. I was feeling strong. My appetite had improved, I was sleeping undisturbed and I was not getting breathless at the slightest hint of physical labour. KP and I started by visiting our three pastures. We identified the walking routes that would be safe from the paths of avalanches which we expected once the big snow came in. We measured the availability of grasses and herbs across the three pastures.

I was also busy making observations of what the blue sheep ate. I wanted to practise watching the blue sheep eat without disturbing them, and I used the tail end of the rutting season to get close to them. I was hoping that they would remember me after the rut and think of me as another harmless person from Tashigang.

Our first snow of the year fell in the third week of December. The dull brown rolling hills of the cold desert were transformed into a white carpet and merged with the big mountains in the background. With it, we lost road connectivity, and electricity was rationed. The bulb in the room would light up only from six in the evening till eleven at night. The rut was over. The big rams lost much of their testosterone and looked tired and dull. They barely moved all day. Sometimes, they would try to nibble a blade of grass from their sitting position. My lungs were working at full capacity, and my legs found their rhythm on the trails leading from our base camp to the three pastures. Sushil had helped me buy half of a *dimo* (a female yak). Sushil, Thinley, KP, the teacher and I butchered and skinned it. All four neighbours came to help and, as per tradition, we shared with each of them a portion of the meat and a portion of the sausages made from the blood and organs. Nothing was wasted. Everything was cut up, packed and stored in a closed room without heating for it to freeze. This meat would supply the protein for KP, the teacher and me for the next five months. I could not help but think that if three people needed so much food for the winter, how much food would a herd of sixty blue sheep require? The teacher offered to pay for his share of the meat but I refused. He insisted and I persisted. Sushil saw it and gave me an approving look. When other families in the village were butchering their meat supply for the winter, we helped and received our share of meat and sausages. We had become part of the network of indebtedness in the village. The elders remarked that the coming of three young people made the winter livelier than the summer. KP and I were out in the field during the day. The teacher would cook dinner for us and many evenings the whole village gathered in our little room (sometimes I wondered how it was even possible) and we sang and danced and drank the Chang beer and arak brewed from locally grown barley.

Two months into my fieldwork, my routine was set. I would take a day off every week. On one of those days, I visited Sushil's house in the neighbouring village, Kibber. I was lazing around on his terrace, sipping tea. The ground was monochrome white and the sky a deep blue. No roads led in or out. I was drifting on a raft in the open sea covered in ice.

A little boy from the village climbed up my wooden ladder to the roof and started peering through the spotting scope which I had set up earlier that day. I had become accustomed to people from the village using my equipment without permission. This lack of formality gave me a sense of being accepted as a part of the place. He saw something and concluded that it was a blue sheep stuck in the snow. I laughed. 'Blue sheep are adapted to living in these conditions and would not get stuck in the snow.' He did not contest me.

But now, my curiosity got the better of me. What could it be that the boy had confused for a blue sheep stuck in the snow? I looked through the scope. What I saw was one of the biggest surprises of my life: a snow leopard. My first sighting in the wild. It was a kilometre away, silently plodding through two feet of powder snow; with only head and back showing, she was swimming in the deep snow. Just as Rajesh had described, the snow leopard's movement and silhouette had given her away.

I kept my spotting scope focused and tried hard to see where the snow leopard was headed. Suddenly, I noticed another movement in the corner of my view. Now I could see two snow leopards walking parallel to each other, 20 feet apart. Enthralled, I called everyone, and then ran downstairs. Moments later, Sushil, Thinley, the teacher, KP and I were headed to a place where we could hide and wait for the approaching snow leopards. It was a short hike to the edge of the gorge that separate Kibber from the neighbouring village called Chichim. The two snow leopards arrived without delay. They were across the deep gorge from us, but the distance as the crow flies was less than 200 metres.

One snow leopard was older than the other. They were a mother and a cub. Before leaving for fieldwork, my supervisor, Charu, had sent me a few pictures from a camera-trapping project which had found four snow leopards. One was a female they had named Sunshine, for her tendency to be photographed on sunny mornings. She had a cub who was rarely photographed and was left unnamed. There was also a large male who had lost his tail, perhaps in a fight or an accident. He had been named Tail-cut. And there was another large male, Eureka, whose name spoke of the scientists' delight when he became the first snow leopard ever camera-trapped in the region. The unique rosette markings help identify individual snow leopards but getting to know their sex is very difficult. You have to wait until to get a peek at the scrotum to know if they are males or wait until you find cubs in tow to know if they are females.

This had to be Sunshine and her cub. While Sunshine lay in

the snow, her cub played with her tail. All of a sudden, both were still and alert. We could feel their tension thicken the rarefied air. Further away, on the same slope, we noticed movement and re-trained our lenses. Another snow leopard. I couldn't believe my eyes. The snow leopard is one of the most elusive wild cats in the world. In the ten years that there had been a research team in Spiti, there had only been one sighting of a solitary snow leopard – and we were watching three together. I knew of scientists and film-makers who had spent years in the Himalaya looking for this beautiful animal and never seen it. Transfixed, we saw Sunshine remain un-perturbed but the cub shrank in fear. If Sunshine was scared then she did not show it. The cub's movements softened; he crouched, belly brushing the floor, almost disappearing into the surroundings. He became a shadow of his mother.

Shadow is what I named him.

The third snow leopard appeared to be a large male. Eureka? He had his tail intact. He stayed about 100 metres from Sunshine and Shadow, hidden in a rock crevasse. The tension persisted for over an hour, throughout which Sunshine, basking on the open sunny slope, kept a close watch on the new male while her cub stayed in a rocky hideout, peeping outside at regular intervals. With his eyes trained in their direction, it was clear that the large male was aware of the cub's presence but he never displayed aggression. In many large cat species, males are known to be aggressive towards cubs that are not their own, sometimes even killing them. But why was Eureka so calm? Was he the cub's father? We weren't able to pick up identifying features during our observations so we couldn't confirm if he was Eureka. We couldn't even confirm if he was one of the resident leopards or a newcomer to the area.

The snow leopards could see us as there wasn't anywhere for us to hide in the open desert-like landscape. But they did not seem to mind us. Perhaps the deep gorge between us gave them a sense of comfort. I was surprised that these mythical and shy creatures did not mind human presence if they could be assured that they would not be disturbed.

Time flew by; it had been over five hours since we had spotted the first leopard. The sun dipped below the ridge in the west, and the temperature plummeted. The thin air of the high altitudes does not hold much warmth. It was time for us to leave. I was unable to hold the binoculars as my fingers were frozen inside my gloves. We left the snow leopards after it became too dark to spot any movement, even against the bright snow. But we kept glancing back, hoping to get one last look.

Back in the village, everyone gathered at Sushil's house, and we shared our observations of the snow leopards. Everyone insisted that we celebrate as if it were a wedding or a birthday. I had to buy a sheep and have a feast for the entire village that same night. Soon, plates of meat dumplings were going around.

I had managed to take a few pictures with my new digital camera, which was being passed around. An old woman came and sat next to me; I could not guess her age but she was the oldest person I had seen in Spiti so far. At first, I could not understand what she said. She spoke Spitian, a dialect of Tibetan, without a hint of Hindi or English. Soon, someone joined us and started translating. The old woman wanted to see my picture of the snow leopard. I was careful to zoom in to show a close-up view of the animal. She had tears in her eyes. She had lived in Kibber her entire life and had never seen a snow leopard. Fewer than a handful of people in the village had had brief glimpses of the snow leopard. It was the first time that anyone had seen them in a long time. The old woman thanked me for sharing this picture with her. She got up and left without another word.

That night, I stayed up thinking about the snow leopards and what the old woman had said to me. My head was full of questions about the snow leopards and the people who lived here. I had read that pastoralists disliked carnivores like snow leopards because they kill their livestock, yet here was a pastoral village celebrating the sighting of three snow leopards. Sushil had brushed aside my question, saying that Spitians only needed a reason to party during the snowed-out months of winter. But the old woman's tears did not fit Sushil's explanation.

And who was the large male snow leopard? Would he attack

the cub in the dark, or would they be fine together? What about Sunshine? Early winter was the breeding season of the snow leopards. Would she mate with this large male? Was her cub old enough to wean and look after himself? We had seen three snow leopards on one day. The camera traps had recorded four earlier in the summer. How many snow leopards were there in the Spiti Valley?

At first light the next day, we were back at the site but the snow leopards were gone. Wind had cleared what little remained to be read in the pug marks. There was no evidence of the presence of the animals from the previous day, just uniform windswept snow all around.

I saw the snow leopard twice more during the remaining five months that I stayed in Spiti. On both occasions, I was alone and I had time to consider how I wanted to go about studying them. Other than the four snow leopards that had been captured on the camera traps, I suspected that there were two more. Six snow leopards in the small area around the Tashigang and Kibber villages were very promising.

My research on the blue sheep showed that the lack of grass availability was indeed driving the blue sheep to feed on herbaceous vegetation, but it came at a cost. In the pastures where livestock did not graze and there was enough grass, 90 per cent of all females had kids with them, but in the intensely grazed pasture with little grass, only 30 per cent of females had kids. We had found a direct connection between blue sheep diet and the performance of their populations. I had not observed any mortality in the grown-up males or females and this was surprising because they seemed to be doing different things. The males were sitting around all day without eating much, just conserving their strength, while the females actively fed throughout the day. Since many of the females were pregnant, they had to keep eating. It seemed that the males and females had different strategies to survive the winter. But that would be a research question for another winter. I packed my bags and headed back to Bangalore to present my research findings. And I went back with something more: a resolve to return to study the snow leopard.

# 3

# Dumplings for Snow Leopards

'So you don't like noodle soup?' Takpa remarked, seeing me reach for the dumplings and ignoring the noodles. Stanzin Takpa had joined KP and me in my new research project in Tashigang, and was both my host and my assistant. Then in his mid-forties, he had an unusually large moustache for someone from this remote mountain region in the Himalaya, and the way he wore his long, straight hair – pulled back with a hairband – also set him apart. This was my second winter in Spiti, having returned a year after my master's field project. As a guest in Takpa's home, I'd sometimes have dinner with him, his wife and two children. During the thick of winter, they withdrew completely from the other areas of the house to a single living room in the basement. There, everyone sat around a rectangular stove kept alight with dried cow-dung cakes. At mealtimes it doubled as a cooking stove and family members and guests would eat huddled around it. After dinner, they would pull out their bedrolls and sleep there too. I had a similar set-up in my spare room upstairs. The only difference was that my room was more exposed to the freezing winds that hurtled past the building and it was much harder to keep warm.

The most common meal in Takpa's house was noodle soup. *Thentuk* is made of barley or wheat flour mixed with water and salt, formed into chunks and then boiled with hard cheese, dried nettle and pieces of meat. But on special occasions, Takpa would make yak-meat dumplings, striking the perfect balance between red meat and fat that made them juicy and satisfying. Served with a tomato chutney mixed with *gemen* (allium seeds) collected from the pastures, this was the best meal you could get in a snowed-in village in the high Himalaya.

'I like noodle soup but I prefer dumplings when they're on the menu,' I said. 'But otherwise, I'll take the soup.' In truth, I really don't like *thentuk*, even if it has chunks of yak meat.

As I spoke, I realized that this was the same logic that was driving our understanding of what snow leopards like to eat. We, the scientists, believed that wild herbivores such as the blue sheep and ibex were the dumplings and that the villagers' livestock, such as sheep, goats, horses and yaks, were noodle soup. We believed that snow leopards would always pick dumplings given the choice, only eating livestock when blue sheep were not available. But snow leopards preying on livestock remains an important conservation challenge. Losing stock is a real problem for herders like Takpa, who are often forced to take matters into their own hands.

In 2005, my mentor Charu won the Whitley Award – known among scientists as the green Oscar – for developing conservation models to resolve the problem of livestock killing by snow leopards. His genius was to implement two solutions: the first was to help the herders receive financial compensation when their livestock was killed by snow leopards; and the second was to increase the populations of wild herbivores like blue sheep so that snow leopards would have enough to eat and livestock predation would reduce.

But there was a problem. If word ever got out that Takpa had made dumplings his neighbours would visit on one pretext or another and stay for dinner. Invariably, there wouldn't be enough dumplings and Takpa would cook more noodle soup to make up the difference. As Takpa said when he roused me from my thoughts: 'If you like dumplings so much then we will make them more often – just don't tell the neighbours.' This problem when considered from the perspective of the snow leopard meant that perhaps the more blue sheep in an area, the higher the numbers of snow leopards – and so livestock killing would be needed to supplement the diets of this higher population.

Charu had started the two programmes in the neighbouring village of Kibber. Kibber is a scenic village, similar in appearance to the movie set of Rivendell from *The Lord of the Rings*, with seventy low houses perched on a rock with a river cutting a gorge around its base. A small stream flows through the village, crashing over rocks into the river below, whilst the 6,000-metre peak of Mount Kanamo looms above. It is a magical place, exactly the sort of landscape where snow leopards *should* dwell. Yet the people of Kibber are more like the warm residents of Bag End in the Shire than immortal elves. When I first visited, I was invited to a party almost every evening.

I had read everything there was to read about Charu's work in the village, but it was at these parties that I came to understand the extent of his work. In 1998, for instance, Charu had worked with the local people to set some of their pastures aside for the blue sheep. The prompt had been Charu's research showing that intense competition for grasses and browse from the livestock of Kibber and neighbouring villages had suppressed the blue sheep population in the region. Kibber had also been renting some of their pastures to nomadic graziers who came to the region during the summer months. Charu convinced the Kibber *yulva* or village assembly to rent the pastures to him instead so that the land could be rested a little and the blue sheep population could bounce back. This would also help the pastures recover from their degraded state and benefit Kibber in the long run. Once some of the pastures were rested, the blue sheep would have enough to eat and their populations would increase.

The plan was to put dumplings back on the menu.

The media had gone on to assert that increasing the population of wild herbivores in this way was a silver bullet that would address all conservation problems. They claimed that an increased wild herbivore population would not only support a healthy population of snow leopards and reduce livestock predation, but it would also help improve the attitudes of local herders towards snow leopards. Charu was less convinced, but he believed in trying all ideas that held promise.

The results would take time. Female blue sheep bear only one offspring every year. If the winters are harsh, the young may not survive. So while the blue sheep population recovered, Charu had helped the village create its own livestock insurance programme. It worked similarly to a life insurance scheme that ensures that our families get the necessary support if something were to happen to us. He convinced the herders in Kibber to contribute a small premium for every animal they owned. This common pool of money was to be managed by a group of people from the village, chosen by the village. And a conservation organization, Nature Conservation Foundation (where I would go on to spend most of my working career) would match the amount. Every time someone lost a goat, a sheep, a yak or a horse, they would be compensated at market price from this common pool of funds.

I remember reading about Charu's Whitley Award in the newspaper when I was still a teenager. He was my hero, and I wanted to be like him. After I completed my master's dissertation, I was hoping that he would accept me as his PhD student. For my research, I wanted to test whether increasing the population of blue sheep really was leading to a reduction in livestock killing by snow leopards. This was one of the fundamental assumptions of Charu's award-winning idea.

But I had grown up in the Indian education system which discourages any criticism or questioning of your mentor. Throughout school and then my undergraduate degree, I had always touched the feet of my teachers before entering an examination hall to seek their blessing for good grades. Questioning my teachers or challenging their ideas was met with punishment. The physical and mental scars of those punishments ran deep.

In 2008, when I met Charu after completing my MSc thesis, he congratulated me and asked what I was considering as my next step. I was quick to say that I wanted to do a PhD but when he asked what I wanted to study, I embarrassed myself by beating around the bush and not saying anything clearly. Thankfully, Charu saw through my hesitation and offered me a job, saying I could also use this time to develop a proposal for my PhD.

My first job was to draft the management plan for the Spiti Wildlife Division in Himachal Pradesh, and so within weeks of completing my master's degree, I was back in Tashigang having dumplings with Takpa. This time I would not be following a single herd of blue sheep through the winter, but travelling the region to understand its ecology and people. I would use Tashigang and Kibber as my base to recover from these trips. The aim was to suggest the best possible way to manage this landscape for wildlife conservation and peoples' livelihoods using some of the new ideas developed by Charu and other senior colleagues.

On the first day of my job, I had driven my 4×4 Maruti Suzuki Gypsy to the top of the Kunzum Pass, at an elevation of 4,500 metres. A high mountain pass which connects the Spiti and Lahaul valleys, it is also a watershed, for behind me was the deep gorge of the Chandra River. This river has carved a narrow valley where it meets the River Bhaga and becomes the mighty Chandrabhaga. From there, it flows into Kashmir where it is known as the Chenab until it enters Pakistan and joins the great Indus. And in front of me was a little stream, the birthplace of the River Spiti. The Spiti flows east, carving a wide valley between two mighty Himalayan ranges, the Parang in the north and the Greater Himalayan in the south. It keeps a steady course and meanders through the high-elevation plateau until it meets the Satluj and cuts sharply through the Greater Himalayan range to emerge on the plains of North India. Eventually, the Satluj River also meets the great Indus in Pakistan. As I stood there, the snowflakes that landed in front of me would have to flow over a thousand kilometres before they met once again with the flakes landing behind me.

Lying between the Zanskar and the Greater Himalayan range, the Spiti Valley nests perfectly between the Himalayan communities that live on the southern slopes of the Greater Himalayan mountains and the Tibetan communities that live to the north across the Zanskar range. For over a thousand years, Spiti was also the staging ground for the trade routes connecting India with Tibet – until 1962, that is, when the war between India and China closed the border and Spiti became a neglected frontier landscape.

As I wandered the surrounding valleys and slopes, I learned that everything in the region depended on the shape of these mountains. The Himalayan ibex, another important prey species for the snow leopard, kept to the steep slopes of the peaks found in the western and southern region of the Spiti Valley, whereas the blue sheep kept to the gentler slopes in the north and the east. Snow leopards were found almost ubiquitously across the entire region, whereas the Tibetan wolf stayed close to the flat areas around the riverbeds. Villages in the rocky areas had more goats as livestock, but those with extensive green pastures had yaks, horses and cow-yak hybrids.

With every passing day, I was learning more about the snow leopard by talking to the locals and hiking the mountains. And every day I found myself thinking about how to design a study to accurately test how increasing the blue sheep population affects the livestock-killing behaviour of this canny predator; and how I would bring myself to discuss my ideas with Charu. For this, I needed to study the prey preference of snow leopards, still little understood; the effect of prey availability on their population; and, finally, if snow leopards would change their prey preference based on what was available on the menu.

Controlled experiments are the gold standards of scientific research. Much of modern science depends on evidence coming from laboratory conditions where everything is controlled by the experimenter, except the effects on the one thing they're studying. Much of the theory of predator–prey relations has also been developed in labs where ecologists would present different prey species to an insect or a fish predator and then study its effects. Over decades, we have built enormous knowledge through these experiments.

When studying a particular animal's food preferences in the lab, scientists often do 'cafeteria experiments'. As the name suggests, the researchers provide animals with multiple food choices and observe what they prefer to eat depending on availability and variety – much as Takpa was observing my own choices in his home. However, conducting such experiments with snow leopards is almost impossible. Snow leopards are hunters and presenting them with choices of wild herbivores versus livestock under the controlled conditions of a zoo is unethical. Even if this were somehow acceptable, the results would not be applicable in the real-world conditions of the Himalaya. The chances of success and the risk of injury for the snow leopard when hunting a blue sheep in the mountains are very hard to replicate.

I knew I had to find a way to study their food preferences in the mountains themselves. I also knew that I could study the diet of the snow leopard by studying their scat samples. Snow leopards, while eating, ingest indigestible hairs of their prey. But how would I know how the diet of the snow leopard changes with a changing menu and changing availability?

One possible approach was to study the diet of the snow leopard across many different places with different populations of wild herbivores, such as the blue sheep, ibex, argali (the largest wild sheep species), as well as livestock such as sheep, goats, yak and horses. By studying snow leopard food choices across various locations, I could observe whether the proportion of livestock in the diet of the snow leopard was reduced in places where there was a higher population of wild herbivores.

Thinking through all of these scenarios was a mental workout, but implementing them across large swathes of the Himalaya was a physical challenge. I had persuaded my mentors to let me do my MSc thesis in the mountains with great difficulty. I suspected that convincing them to let me study snow leopards in multiple locations across this vast mountain range was going to be a new challenge.

I steeled myself before speaking to Charu, yet his response was pleasantly surprising. I was accepted as a PhD student, supervised by Charu at NCF with co-supervisors Yash Veer and Professor Steve Redpath from the University of Aberdeen. The scepticism that I was worried about when presenting the seminar for my master's thesis wasn't present among these three men. Studying snow leopard diets across multiple sites was the easier part, they all agreed. The hard part was to find out what was on the menu at each site. Nobody had successfully estimated the populations of wild herbivores in the Himalaya over large enough study areas. I would have to come up with a method to count the wild herbivore population before embarking on the exploration across multiple places. One huge study was required, to make another huge study possible.

Studying the populations of wild herbivores is harder than it seems. To know the livestock populations, I could simply talk to all the herders in a given place. But to do the same for wild herbivores is a different challenge. It is easier to estimate numbers of the big cat species. As tigers are striped and snow leopards have rosettes in individual patterns, their markings are like fingerprints that allow scientists to identify and count individuals. But the blue sheep and ibex don't have such markings and can't be counted the same way.

In the forests or on the plains of Africa and India, scientists use something called the Distance Sampling Method to estimate herbivore populations, but this involves walking in a straight line along a randomly chosen bearing across large distances while counting herbivores and measuring their distance from the observer. In the Himalaya one cannot walk more than a few metres in a straight line before falling off a cliff. So, distance sampling was not an option.

I read widely to see how others had tackled the problem. Scientists from the Denver Zoo had flown in aeroplanes over the Altai mountains in Mongolia and successfully counted the population of wild herbivores. Although tempting, it would be too expensive and I did not fancy low-altitude flying over the Himalaya. I am a mountaineer who prefers to have my feet firmly anchored on the slopes. Being in the air with little control over the wind currents scares me more.

Finally I came across a little-known research paper from the *New Zealand Journal of Ecology*. Researchers there had developed a method to estimate the population of the Himalayan tahr – a large and primitive species of mountain goat with a thick, reddish coat. Historically, the Southern Alps of New Zealand do not have any non-marsupial mammals, but in 1904 the Himalayan tahr was introduced for sports hunting. As with most colonial-era introductions of animals, this has been an ecological disaster. New Zealand has some very rare and endemic plants and birds, and the tahr love to feed on those plants, simultaneously destroying the birds' habitats.

Over time the Himalayan tahr did so well that its New Zealand population far outnumbered its population in the Himalaya. The International Union for Conservation of Nature (IUCN) estimates that there are no more than 10,000 tahrs across the entire Himalaya, but in New Zealand an aerial survey in 2022 estimated the population to be over 34,000. The authors of the paper I'd found had done something remarkable, however. They had used one of the oldest statistical techniques conceived – the Double Observer Survey – and applied it in the mountains of New Zealand to accurately estimate the population of the tahr way back in 1997.

I saw hope in the Double Observer Survey method. I wanted to adapt it for my research context and deploy it to estimate the population of herbivores.

'Are we counting them twice to confirm that we counted correctly?' Thinley asked when I tried to explain the Double Observer Survey method. Thinley was an excellent naturalist and a born leader. By now we had been working together for two years. He was already leading NCF's team in Spiti and fifteen years later at the writing of this book, the two of us work together still.

The thought that someone might ask me this had crossed my mind many times and yet I found myself unprepared.

'No, Thinley-ji, we are using a statistical method that requires us to collect the data in this manner. We are counting them twice for reasons to do with statistics.'

I knew I made no sense.

'Look, I know and understand that there must be important reasons why we are doing it this way,' Thinley said, with his wise-man expression, 'but your silence on this matter and your evasive behaviour is raising suspicion within the team. You need to talk to them.'

This was crucial advice. The team suspected that we were counting the blue sheep twice because I did not trust them to do the job properly. My team of eight comprised local herders and farmers who had come together to help estimate the population of blue sheep in one section of the Spiti Valley. I knew from my mountaineering days that trust between team members was more important than the skills that we possessed individually. But I had no clue how to explain without sounding pedantic.

Thinley-ji sensed my dilemma. 'Don't worry about it, we don't have to talk through every detail, just help everyone see the logic of what you are trying to do.'

As the sun went down on the first day in the field, we had an early dinner. All eight of us were sharing a room – at least we would be warm. As I was preparing to get into my sleeping bag, the deck of cards came out. Sleep deferred, we began playing *teen-patti*, a form of three-card poker. I was losing every game when it struck me. I interrupted the proceedings, gathering together the cards and set some aside, uncounted. Not understanding what I was doing, Thinley called me a bad loser. In response, I asked him to guess how many cards I had in my hand. He blurted out a number, irritated.

'Imagine,' I replied, 'that the unknown number of cards in my hand are the unknown number of blue sheep in the field. We have to estimate this number because we cannot actually count all the blue sheep.'

I asked Thinley to look at the top five cards in my hand. Then I took them back and shuffled the unknown number of cards. I called him to look at the top five cards a second time. Two of the cards were the same as those he'd seen during his earlier draw. I could see the card player in him estimate the chance of drawing the same card twice. If the number of cards in my hand were very large, then this chance would be small. If the number of cards in my hand were small then this chance was higher.

As he did so, I saw a twinkle in his eyes. He had figured out what I was trying to do. We turned towards everyone else and said it almost together: a Double Observer Survey is like counting cards to get at the probability of having missed some the first time around. 'This is what we will try to do with the blue sheep herds. If the number of herds in the field is small then the chances of seeing the same herd twice are large and if there are many herds then the chances of seeing the same herd twice are small.' The team looked back at me, unimpressed. Yet they saw the logic. And we all went back to playing cards.

Just like that, I was one step closer to my research project. In a couple of weeks, our field team surveyed a few hundred square kilometres of snow leopard habitat to estimate the blue sheep population using the Double Observer Survey method. That done, I now needed to know how to estimate the snow leopard population.

There are two main methods of estimating the population of snow leopards within a region. The more popular method is to set up automatic cameras that are triggered by movement and thermal sensors. These cameras photograph everything that steps in front of them. We can then use the unique markings on the snow leopards' coats to identify individuals and count their numbers. The other method relies on scat, with scientists collecting samples from a region and extracting the DNA of the snow leopard back in the lab. Since all animals have unique DNA, we can then count them. In both approaches, the hard part is to be sure that we don't miss any animals, either because we don't get their pictures or because we did not come across their droppings. The analysis was based on the theory of probability in either case. And although setting up automatic cameras in the field sounded like fun, I decided to use the DNA method. I would be collecting scat samples anyway to study the snow leopards' diet; the same samples could be used to estimate the snow leopard population.

My grand idea was to find places in the Himalaya and other mountains of Central Asia that had different prey available for the snow leopard. At one extreme, I needed a site which had a large livestock population and very few wild herbivores, and at the other I needed a site that had a lot of wild herbivores and very little livestock. Then a bunch of places in the middle. Each of these sites had to be at least 300 square kilometres large – the same size as the island of Malta. All this fieldwork would need money and I wrote to every funding agency possible. My student stipend was being paid by the Nature Conservation Foundation but funds for fieldwork were slow to come until I was awarded the National Geographic Young Explorer grant. As I got better at writing grant proposals, small grants began to arrive and I hit the mountains.

Over the next two years we established these sites, conducting Double Observer Surveys to estimate the wild herbivore populations and speaking to local herders to conduct a census of the livestock population. Using these methods, I identified seven sites. Given my familiarity with the Spiti Valley, five were there. I also chose one site in the Zanskar mountains of Ladakh in eastern Kashmir and another in the Tost mountains of Mongolia. Across all these places, people were herding almost the same types of livestock – goat, sheep, horses, yaks, cows. And there were different types of wild herbivores, all of which were part of the snow leopards' diet. The blue sheep and ibex were the most widespread. Some sites had argali; others had urial, the wild ancestor of the domestic sheep.

I was doing fieldwork for this study in the Tost mountains in the Gobi Desert of Mongolia in 2011. I was walking along a ridge leading to the highest peak of the Tost mountains. The drop on either side was gradual until it disappeared into a mass of broken rocks at the base of the mountains. In the distance, these rocks merged into the sands of the Gobi Desert. It was October and there was not a single blade of green grass. Even from my high vantage point, I couldn't see the horizon. The dust of the desert simply blended into the grey sky. The terrain was as different as it could get from the familiar Himalaya and yet snow leopards lived in these mountains with as much comfort as they did there.

I was walking the ridges because that is what the snow leopards like to do. Up here, they scan the mountainsides for herds of ibex, a potential next meal. As I walked I paid attention to every little pebble and the larger boulders in my path, looking for any hint that a snow leopard had come this way over the past few days. The ridge told me the story of these mountains. The slopes descending from the ridge were gentle, and the rocks loose, suggesting that these were very old ranges. The weathering and erosion of millions of years had rounded most of the rocks, even hollowing some of them out – a sign that dry winds were the main eroding force. The top of the ridge was not a knife-edge, as in the young Himalaya. Here it was blunt and well worn. Animals and people had walked these places for thousands of years, creating what was almost a path. Tiny hoof prints of goats and the little black dots of their pellets indicated that the local herders came here often. The occasional large hoof print suggested ibex did too. As I came upon a saddle in the ridge, my expectations grew. Saddles are often the places where snow leopards like to mark their territory by defecating, urinating and making scrape marks in the loose gravel, little heart-shaped depressions made with their hind legs.

As I reached the saddle I immediately noticed a fresh print in the soft sand – my first glimpse of a sign left by the snow leopards of the Gobi Desert. It felt good to have my natural history skills validated on new ground. The pug mark was so fresh that the snow leopard was probably walking only a little ahead of me. I was about to push on, to scan the surrounding area using my binoculars, when something caught my eye. A dark, steaming dollop of snow leopard turd, as if the snow leopard had left it just for me, delivered according to my needs. It was my first sample of snow leopard scat from Mongolia, but the sun would be setting soon. I didn't have enough time to collect the sample with all my equipment and gear, look for the snow leopard and get back to camp safely in this remote part of the world.

The chances of seeing a snow leopard are always very slim, so when the opportunity arises, you grab it. I had only seen a handful of snow leopards in my research so far: thrice during fieldwork for the master's thesis and twice since the start of my PhD. I decided to mark the location of the turd with a handheld GPS and return to it the next day. Right now, I had to find the snow leopard.

GPS in place, I almost ran up the next section of the ridge. My plan was to get to the nearest high point – a large boulder – to try and spot the cat. Adrenalin was rushing in my veins, my heart thumping. As I reached the top of the ridge, eyes already looking for a silhouette of a snow leopard on the surrounding hills, I rounded the boulder.

There was the snow leopard, so close I almost bumped into it. I fell back onto my rucksack, stunned.

The leopard was crouching low, almost flat on the ground as if hoping to get by unnoticed, looking up at me with suspicious eyes. Less than 5 metres away, it was prepared for this encounter, as it must have heard me coming. Up close, its body looked smaller and its tail longer than expected. Without blinking, it turned around and moved rapidly down the slope, gliding through the loose, broken rocks with the seamlessness of a wave of seawater retreating from a rocky shore. It took me a few moments to gather my senses and get back to my feet. By then, the snow leopard had merged with the ocean of rock and sand.

Seeing a snow leopard in a new mountain range with its own unique challenges, while working alone, felt like climbing a new route on a new mountain. It demanded all the same responses: of needing to understand the mountain, its weather, its moods and then joining the dots to overcome a challenge. Finding this snow leopard meant I had solved a new problem; a new puzzle of the mountains was unlocked. It is the unpredictability that makes this search for the snow leopard so utterly exciting. Content with the emptiness and solitude of the Gobi Desert, I went back to my dollop of snow leopard turd. There was time after all.

Two months later, I sat in the genetics lab of the Institute of Biology at the Mongolian Academy of Sciences in Ulaanbataar. I had that very snow leopard scat in front of me in a blue plastic vial labelled TOST-001.

Over the next few days I unravelled the life of my snow leopard, a female. She had eaten an ibex in the days before our encounter. Later, I came across at least four more samples of her scat and every time it was an ibex she'd brought down. I had estimated the population of argali and ibex using the Double Observer Survey in this region and found them to be supporting five snow leopards.

Finally, after two years of living out of a backpack and wandering across these seven sites, collecting hundreds of scat samples, and obsessing over statistics like a gambler over blackjack, I had a story to tell.

The site with the smallest number of wild herbivores had the fewest snow leopards – just one lone beast – and the site with the highest numbers of wild herbivores had the highest number of snow leopards. Kibber and the surrounding areas were very well populated, with eight – two more than my guess at the end of my master's study. Across seven sites, I had estimated thirty-three different snow leopards. To know that I had been observing and tracking that many snow leopards was intensely satisfying.

My study showed that a lot of what we knew about the snow leopard's diet was correct. Snow leopards liked eating wild herbivores more than they liked eating livestock, despite the fact that hunting down an ibex or a blue sheep is much harder than hunting a cow or a goat. Perhaps the snow leopards were also weighing up the risk of retaliation by the herders. And because snow leopards preferred to eat ibex and blue sheep, increasing the population of these wild herbivores would lead to every individual snow leopard killing fewer livestock.

However, I also discovered, as suspected, that the neighbours are always watching. Increasing the population of wild herbivores led to an increase in the snow leopard population, either because they were breeding faster or because snow leopards from neighbouring areas were moving into areas with more wild prey.

Even when each individual snow leopard killed fewer livestock, when more snow leopards were around, together their kills added up to more livestock being killed. More wild herbivores meant more snow leopards – success of conservation – but it also created a new challenge for conservationists to solve. Takpa was right. If I wanted more dumplings, I had to ensure that the neighbours didn't find out.

In the process of this research, I learnt an important lesson: ecology is a complex web of interactions between different species, including humans, and there are no silver bullets for achieving conservation outcomes. Perturbations to the web in one place will have consequences in multiple unforeseen ways. Conservation is also about managing the consequences of conservation actions.

I spent two sleepless nights before breaking this story to Charu. I was worried it would seem that I was finding faults in the conservation model for which he had deservedly earned recognition. My life in the Indian patriarchy had taught me to be extremely cautious when challenging the set notions of my seniors. All too aware that I was only 26 years old and Charu was one of the most respected conservation scientists in the country, I thought of every possible way of sugar-coating the message.

When I met Charu, I was nervous, beating around the bush, as always. Charu, as usual, saw through my nonsense and asked me to explain the results in detail. He was delighted. He pointed out that while this research had exploded the belief that increasing the wild herbivore population is a one-stop solution, it had also strengthened the idea that wild herbivores are critical to the success of snow leopard conservation. They are indeed the preferred prey and their populations are crucial for supporting snow leopard populations. Yes, more snow leopards meant that between them they would kill more livestock, even if each snow leopard killed less livestock individually, but this could be managed through better protection and compensation schemes.

The team and I concluded this study with a sense of relief and achievement, but also sombreness. In my youthful excitement, I had thought that I was going to change the world with my path-breaking research. Instead, I had shown that the problem of livestock killing by snow leopards was here to stay. All we could do was manage the consequences. I had my story of the predation ecology of snow leopards, but whether local livestock herders chose to live with snow leopards would depend on their own stories of conflict and coexistence with this majestic cat.

# PART II

# People

# 4

# Shen

One morning, towards the end of my dumpling study, Takpa and I were busy looking out for blue sheep. We had walked a few kilometres from Tashigang when we saw a dead blue sheep lying in the open, a mid-sized male who must have been around 50 kilograms when alive. It had a big laceration at its throat, and the abdomen was open. Part of the inner thigh had been carved out. The snow around the dead sheep was pocked in snow leopard pug marks, but there was hardly any blood on the ground. Locals believe that snow leopards drink the blood of their prey and sometimes get high from it. Snow leopards often kill their prey by attacking the throat and quite likely lick the blood, but there is no evidence that they get intoxicated from it. The myth stems from the fact that a snow leopard defending its kill is far bolder than usual.

No vultures, foxes or crows were nearby, a clear indication that the cat had been here only recently. The carcass was barely eaten, so we knew that the leopard would return. We decided to collect as much data as possible and then return in the evening to hide nearby to try to see the leopard as it returned to claim its prey.

This was Sunshine's territory and I had seen her cub Shadow nearby a few months ago. I wondered if it was Sunshine or if Shadow was hunting by himself. He was a young male and there was a good chance that he would leave his mother's territory. I was eager to know more.

We found a small stone compound built as a temporary hold for livestock, about 50 metres from the dead blue sheep on a rolling slope of caragana bushes. The deep-green hues broke the monotony of the landscape and provided perfect cover. It was barely 4 p.m., and the sun was still out. I was expecting a slow evening, presuming that the leopard would not return until after sunset, so I was a little surprised when Takpa pointed towards the carcass through a gap in the wall. I looked through my spotting scope, which was already set in place. At first, I saw nothing, but then three shapes conjured up out of the grey rock and dry bushes. A large mother and two cubs, barely six months old. They moved like smoke, drifting effortlessly in the low light of the evening. It took me a few moments to recognize the familiar shapes and follow them along. They were in a playful mood and every time one of the cubs jumped on their mother, their shapes would distort like the mixing of watercolours.

When my excitement ebbed, I put the dots together. This was Sunshine with a new litter of cubs, Shadow's younger siblings, which Sunshine must have raised through the winter. I wondered if the father of these cubs was the large male that I had seen. My previous study had found eight snow leopards in the region so it could actually be any male from among those. And these new cubs were a clear sign that Shadow had dispersed in search of a territory he could claim as his own. I knew I would never see him again.

Snow leopards have between one and three cubs in each litter. In rare cases there could be four or possibly even five. Unfortunately, over half of them perish within the first year of their life. This is true even for the cubs born in zoos.

Snow leopards mate during the height of winter. Most recorded matings are between January and March and the cubs are born after three months of gestation between April and July. This is also the time when the herbivores are at their weakest and a new crop of wild goat kids and sheep lambs are about to be born. The mother snow leopard's job is therefore a tiny bit easier as she attempts to feed the growing young in her den. A female snow leopard born in spring will typically mate during her fourth winter, when she is 3 years and 9 months old. In rare cases she may also mate in her third winter aged just 2 years and 9 months.

Sunshine and her cubs gorged on the meal before the cubs decided to play with their mother's tail as she continued eating. Sunshine did not seem annoyed in the least. She was focused on her food while the cubs took turns to attack and tumble with the black pompom of her thick tail. Sunshine had a hard year ahead. She would have to hunt more often now that she had two additional mouths to feed. All the kills would have to be near because the small cubs wouldn't be able to travel far. She would have to protect the young ones from other predators like wolves and packs of village dogs. She had a good territory with a high density of blue sheep and I felt hopeful for her and her cubs. She was an experienced mother, but successfully raising two cubs in the bleak Trans-Himalayan landscape would be challenging. Motherhood can be difficult, even for snow leopards.

We watched until our eyes couldn't strain any more. Sunshine and her cubs disappeared the same way they had appeared: out of nowhere.

'You married?' asked the woman sitting next to me in a cable car suspended a hundred metres above a swift-flowing mountain stream in the Spiti Valley.

She could not have been much older than me. I was used to the women and men of Spiti being forthright, but asking my marital status at the start of a conversation was a first. Surrounded by snow-clad mountains, deep gorges and the crisp cold air of the Himalaya, this was also quite the setting. The 4-foot by 3-foot basket suspended under a metal cable was the best transport connecting Kibber with the neighbouring village Chichim. People sitting inside the basket pulled at an anchor rope to move it from one end of the gorge to the other. I had done my share of pulling the anchor rope to move the basket along and now another young man had taken over. It would be fifteen minutes before we reached the far side of the river, so there was no way of evading her curiosity. But before I could think of a reply, she continued, 'Why don't you marry one of your snow leopards?' Her words had a bite. Everyone here knew that I was studying snow leopards. Her tone was light, so it was up to me to brush it aside or to engage her in a serious conversation.

This was a watershed moment in my interaction with the people of Spiti. Until now, I had followed local customs and reserved all serious talk about wildlife and conservation to the respected elderly men in the village. Here, in the most unusual of places, I had been called to interact with another section of society to which I usually had no access. It had never been a conscious thought not to speak to Spiti's women; it was the way society was organized here.

'Why, do you dislike me or the snow leopards?'

She quickly apologized and said that she didn't mean it. She was tired from not being able to sleep for the past few nights because a snow leopard had been frequenting their livestock pen and she and her children had had to stay up to scare it away. She told me how, when her eldest son was little, a snow leopard had killed their only cow and there was no milk for the child for several months until her husband could buy another. She had known about and even benefited from the government compensation and the livestock insurance that NCF, our organization, had helped set up. She was grateful for this support, but added that it did not help her get a night of peaceful sleep. Her husband kept the money that they received as compensation for the livestock that they lost to snow leopards and wolves. Sometimes he used it to buy replacement animals, but other times he just spent it. Yet looking after the animals during winter was a woman's job. The stress of caring for them during the cold months, only to see them taken by the carnivores, was harsh.

I remembered the old woman who had tears in her eye when she saw the picture of Sunshine and Shadow that night after our first sighting. Her response to snow leopards was very different from the woman in the cable car, but both of these women gave perhaps more honest reactions to meeting a snow leopard biologist. The men I interacted with were friends or friends' friends. Conversations with them were full of stories of adventure and excitement. We talked about the treks we had made, the places we had visited and the animals we had seen. Some of these were true but some were clearly fabricated. We had a good laugh when a man from Kibber claimed he had seen a 'herd' of eleven snow leopards. When Sushil and KP poked fun, he said that he was on a bus and the entire busload of people saw it, but he could not remember a single other person from the bus to verify his account. There is a lot of one-upmanship when speaking to a group of men.

I had already spent a couple of winters in Spiti and I had mixed feelings about how the local people perceived snow leopards and other carnivores. On the one hand, I knew that people lost a lot of livestock to snow leopards and many resorted to desperate measures. Charu had published a seminal paper showing that such costs of livestock predation may even be keeping people below the poverty line. On the other hand, I had met people who wanted to see the snow leopard and considered them part of their natural heritage. However, other than anecdotes, I had nothing concrete to go on.

It was around this time, towards the end of my PhD research in 2011, that our team was changing. KP eloped with his childhood sweetheart. His bride's father knew that we had encouraged the couple and helped them escape and so he screamed at KP's friends and called us a pack of wolves.

'She is the most beautiful girl in Spiti and men with big cars were asking me for her hand. And now you made her run away with a good-for-nothing guy who works for an NGO and looks for snow leopard poop all day. You are like wolves that eat their prey alive.' Secretly, we were thrilled to be called wolves, but this was another indication of what people thought about carnivores, carnivore researchers and careers with NGOs.

KP and his bride were hiding at a distant aunt's place in Kaza when I went to meet them. KP was the happiest I had ever known him. Just when I was leaving, he stepped out of the house and spoke to me alone. He said this was the end of the road for him. Their families were negotiating about the couple's future, and they wanted him to take up the job of a mail runner for the Post Office in Kibber. It was a steady job with the government, it paid okay, and he wouldn't have to travel much further than bringing the mail from Kaza three times a week. I knew we would retain our friendship and keep meeting in Kibber, but KP was a great naturalist, and he could climb like a goat on steroids. But he was starting a new chapter in life, and I was there to support him through it. And so another conservationist dropped out because of societal pressures.

I was still trying to make sense of how people in Spiti perceived and thought about the snow leopard. Within a matter of weeks, I had been asked to marry a snow leopard and then my friends and I had been called wolves. I was reminded of a day when the teacher had accompanied KP and me to the field. We left camp before sunrise and planned to get back before noon. These were the sunny days of late winter and the snowmelt was triggering avalanches and rockfalls in the afternoons. The teacher forgot to bring his goggles. When he realised, we suggested that he go back to the camp and bring them but he insisted that he would be fine. By the time we got back to camp at noon, the teacher was rubbing his eyes violently. By evening, he could barely open them.

Snow blindness is caused when the sun's rays reflect off the snow, causing a burning sensation in the eye. One is especially susceptible to it when high in the mountains because of the brighter sun and higher UV radiation. It feels like there is something in the eye, a foreign particle, forcing you to rub it constantly, causing intense discomfort. Usually, it goes away by the second or third day leaving no permanent damage.

We were feeling terribly sorry for the teacher. We cooked an early dinner and sat around the warm stove, eating quietly. The teacher had cotton tapes dipped in cold water wrapped around his eyes (to stop him from rubbing them) and he was eating from his plate by touch and feel.

For the first time, the camp was not filled with laughter and jokes. All we could hear was the cracking of wood inside the stove, and suddenly the dead of the night was pierced by the deep howl of a wolf. Our spoons froze in mid-air. Within seconds, more wolves joined the chorus. They were not far from the camp. KP and I dropped our plates, grabbed our binoculars and rushed out of the camp, heading up a rickety wooden ladder for the roof. We told the teacher to stay and wait.

The night was serene. The landscape was covered in white powdered snow. In the waxing moonlight, the summit of Chau

Chau Kang Nilda was like a bride in white with a wide hemline to her dress made of snow. In the soft glow, the rolling hills looked like sand dunes at dusk. While we were wondering where to look, another round of howls reverberated through the cold night. The moonlight reflecting off the snow was bright enough for us to read a book. We were straining our eyes to pick out details in the distance when four large wolves crested a small hillock close to the camp. We did not need our binoculars to see them. They stood there for a second. We anticipated another round of howling but they kept looking east, away from us, and started running downhill with purpose. We saw them a few more times when they emerged over distant rolls of the land and then disappeared into the horizon.

KP and I narrated the episode in great detail to the teacher and he smiled for the first time all evening.

'Good omen,' he said, sitting cross-legged and blindfolded with ragged field clothes and the certainty of a zen monk.

The teacher's perception of the wolf was very different from that of KP's father-in-law's. Understanding local peoples' attitudes towards carnivores is fundamental to assessing the resilience of con-servation. I wanted to know what people thought of snow leopards and wolves. If a large conservation project were to be proposed in the landscape, would they rally in support or march in protest? My lived experience suggested that there would be a mix of both, but who would support and who would oppose the snow leopards? And what could we do to win over the opposition? The more I spoke to the people, the more complex became the response. I was also unsure whether people were telling me the truth or sharing stories they thought I wanted to hear. There was a bias in whom I was meeting. I was more likely to meet people who were already sym-pathetic to the snow leopard. I needed to more systematically study people's perceptions and attitudes towards snow leopards.

Studying how people feel about a species can be challenging because our feelings are not fixed and measurable entities. Also, different academic disciplines study this same question in different ways. I was trained as an ecologist, someone who studies interac-tions between species. Investigating people's attitudes seemed like

I was encroaching on the domain of social sciences. However, my curiosity was driven by a need to understand something critical to the conservation of the snow leopard.

With advice and support from my PhD guides, I decided to blend my training in ecology with some of my reading in psychology and sociology to design a study where the team and I would ask people about their experiences, beliefs and feelings towards the snow leopard. To avoid the bias of asking people who were similar to me and my colleagues in demographic, we would choose our participants randomly by visiting only every third house in a village. My questions had to be tailored to capture the nuances of the relationship between people and snow leopards while also being relatively brief. Contrary to urban belief, herders and farmers are very busy people who rarely have hours to chat. Before you know it, their animals start wandering into places that may be too dangerous for them. Farmers have a long list of things they need to achieve every day to have a healthy crop. To give ourselves some time to connect to the people we were talking to, we decided to do this study in winter when there are no farming activities, and when herders keep their animals at home.

My plan was to convert their answers into numbers and use statistical models to make sense of these numbers. This statistical ability was both my strength and weakness. Numbers gave me a sense of objectivity and comfort, but when thinking about people's feelings in numbers, it is hard to understand the nuances. My friends at NCF were my guinea pigs. I would ask one of them, 'Would you support snow leopard conservation in Spiti?' I would score a one if they answered in the affirmative. And if they were not sure, I would score a zero. If they responded they were unlikely to support it, then I would score a minus one. Similarly, I would ask, 'What do you think should be done if a snow leopard kills livestock?' I would rate the answers on a numerical scale based on the severity. When I added these numbers across questions, the final number gave me an abstract representation of the things people told me about their lived experiences, feelings, attitudes and beliefs towards snow leopards. On their own, the numbers would mean nothing, but when taken in relation to the numbers for other people in the same village or people from other villages, and for people with different life experiences, they would start making sense. These numbers would tell me how people thought about the snow leopard in comparison with how they thought about the wolf – a reference group.

It was autumn of 2011 and I started a pilot of my study before plunging fully into it during the winter. I chose a village within Spiti Valley that was far away from Tashigang and Kibber, one where I wasn't known. Thinley accompanied me to act as a translator if required. I had a strange feeling when I counted the third house from the entrance to the village and approached its door. It was like approaching a cliff without knowing what was below.

I knocked and said *'julley'*, meaning 'hello', in the most cheerful voice possible. A middle-aged woman opened the door and smiled at Thinley and me. I said that I was a university student doing a survey to explore how people relate to wild animals in the region and Thinley introduced himself as someone from Kibber who was helping me. She invited us inside for a cup of tea. We climbed a dark staircase to the upper floor. This was a typical Spitian mud house with two floors. The upper floor was very well lit by the large windows. The living room was large with a metal wood-burning stove in the centre and low seating on three sides around it. The sitting mats were covered by carpets and lined by a low *chokse* (like a coffee table) with a beautifully carved dragon on the side. I sat on one side of the stove and our host immediately said that I should sit further up, meaning all the way at the head of the sitting area behind the stove, the place for the most respected person in the room. I hesitated, and the host insisted. Very swiftly, she put a small log in the stove and re-kindled the fire. Then she placed three bowls in front of us. One with roasted barley, one with nuts and in the last cup, salted butter tea. Then she put a teapot on the cooking stove which was in a corner of the same room. Thinley offered to help. He said, 'I will make tea while you and Kullu talk about work.' She accepted Thinley's offer after a small protest.

I started with some icebreaker questions by asking her name and about her family members. Slowly, I moved on to the questions about the family's livestock and agriculture. Our host was a very confident woman and responded with a lot of details. Thinley would clarify my question or her answer for accuracy. Before long, we had moved into what felt like a normal conversation between two acquaintances. I had wanted to have this feeling before asking about wild animals. Thinley poured three cups of sweet milky tea and joined us around the stove. I could not have asked for a better start to this new study.

I showed her a picture of a wolf and asked if she knew the animal. She looked at it for a split second and immediately said, 'Of course, it's a *shingu*, like a dog that lives in packs in the jungle and howls at night.' Thinley and I shared a quick glance as we recognized that our host knew what she was talking about. We were happy with how things were going so far. I proceeded with questions about wolves, asking if she had seen a wolf around her village, if the wolves ever killed their animals, and what we should do if the wolves took to killing livestock. Our host answered confidently.

Then I showed a picture of a snow leopard and asked her if she knew the animal. She took the picture in her hands and said it was a *shen*. Still holding the picture and intently looking at it, she said it was a beautiful animal. When I asked if she had ever seen one, she said she knew that the animal lives around the village and sometimes it gets into livestock corrals and people scare it away, but she had never seen one with her own eyes. She repeated that it was a beautiful animal and handed back the picture. When I asked what we should do when snow leopards kill livestock, she replied, 'It is a problem when they get into our corrals and kill many sheep and goats in one night. But it is fine when they kill the weak and injured goats in the pastures. After all, snow leopards need to eat too.'

My pen stopped moving for a second before I wrote down that sentence. I had lived in Spiti for a few years now and knew the local people to be some of the warmest and friendliest anywhere. But I had never heard or read about pastoral people saying that they accepted carnivores needed to take a few domestic animals in order to eat.

Our host went on to tell us a story of how she had raised a yak calf when its mother died of an unknown disease within weeks of giving birth. The calf would look at her the way a child looks at her mother. She had tears in her eyes when she said that one day someone from the village told her that a wild animal had killed a young yak. She stayed up all night praying that it was not her yak, the one she had raised from birth. She told us how she broke down when her husband confirmed that it was indeed her yak. She was sad for many days after that incident.

There was a lump in my throat, and I felt tears welling up. When we were ready to leave, the host said that we should stay for lunch. Thinley courteously explained that we had to talk to many people and we needed to be on our way. She said we should come back when we were hungry and done with all the talking, and sent us off with a *'julley'*.

We skipped the next two houses and visited the third.

Around this time, our team in Spiti was growing. New students were joining and one of them, Saloni Bhatia, had published an important research paper analysing how mainstream mass media reported incidents of common leopards in urban and suburban areas of Mumbai – one of the world's most densely populated cities. Her research showed how the words used by newspapers can influence public attitudes towards leopards.

The common leopard is a distant cousin of the snow leopard, both of whom are part of the cat or Felidae family.

The snow leopard is genetically the closest relative of the tiger, but because it looks and acts more like the common leopard, it is often confused with it. Among all the big cats, common leopards have the widest distribution. They are found all the way from the Amur region of far-east Russia, through China, South East Asia, India, Sri Lanka, the Middle East and over much of Africa, except for the Sahara Desert. Unlike the snow leopard, the common leopard shows great flexibility in adapting to its surrounding environments, ranging widely in size and appearance.

The Indian peninsula is one of the strongholds of the leopard.

They occur in every possible habitat from the rainforests of the western ghats, to the grasslands at the foothills of the Himalaya. More importantly, they are well adapted to living alongside humans in urban and rural areas, with their stealth and camouflage making them almost impossible to detect. But not every interaction is peaceful. The legendary hunter turned conservationist Jim Corbett shot two famous leopards which had proved particularly dangerous to humans: the 'Panar Leopard' is estimated to have killed around 400 people and the more famous 'Man-eating leopard of Rudraprayag' is reported to have killed around 125 people. Even to this day, nearly a hundred people die in leopard attacks each year in India.

The snow leopards have a very different record. Across their distribution range, from Russia to India, there is not a single documented case of a snow leopard attacking a human. Not even a child, ever.

That does not mean they are not capable. In my studies on the diet of snow leopards, I found horses to be one of their favourites. When Thinley and I examined some of the horse predation cases in Kibber, we realized that snow leopards were scaring the horses to run into ravines with deep snow where they would sink up to their knees or sometimes up to their belly, making it hard for the horses to kick or run. The snow leopards, with their wide snowshoe-like paws, glided over the snow and got the horses by their throats. I knew of snow leopards bringing down fully grown yaks, especially in rocky terrain where the yaks found it difficult to keep their footing. In one case in Mongolia, I knew of a snow leopard that killed a Bactrian camel, which is likely to have weighed over 500 kilograms. Unfortunately for the camel, it was tethered by its owner, which made things easier for the snow leopard.

The bulk of snow leopard prey species like the blue sheep, ibex and argali all weigh more than the snow leopards. Their ability to bring down large prey is further remarkable because they don't have the strongest bite force for a cat of their size. Sister species like the tiger and lion have more than five times the bite force and even common leopards and puma have nearly twice the bite force. Snow leopard bite force is only two times that of a red fox which primarily takes only hares and rodents. Whatever snow leopards lack in their ability to bite, they make up for by using the treacherous mountain landscape to their advantage.

I have often wondered about what goes through the mind of a 35 or 40 kilogram snow leopard when it decides to take on a 400 kilogram horse or a 500 kilogram yak. But the challenge of hunting a blue sheep or an ibex is not any simpler. Over the years, snow leopard researchers, film-makers and tourists have amassed amazing footage of snow leopards falling off cliffs as high as 70 feet and tumbling for further tens or hundreds of feet when trying to hunt wild sheep and goat. In a particularly thrilling video from Kibber, a snow leopard is seen tumbling over rocks, snow and ice for over 200 feet of a vertical precipice, all the while holding on to the throat of a blue sheep larger than himself. When they finally land at the bottom of the valley in a narrow ravine, the blue sheep is stone dead but the snow leopard shakes off the snow from its coat before a pre-meal leisurely stroll. In most such incidences the snow leopards walk away with perhaps nothing more than a scratch. But then Thinley and our team from Spiti have also found dead snow leopards buried in avalanche debris, and once together with a herd of ibex. We considered the scenario where the snow leopard's chase of the ibex triggered the avalanche killing them all. The snow leopards use all their nine cat lives to hunt their prey. They do not lack the strength, stealth, skill or the courage to attack humans and yet they decide to shy away. The snow leopard is the only species of big cat with this noble record of never having harmed a human in recorded history, and I hope it stays this way.

Saloni had a lot of experience of talking to people about the common leopard but she was new to Spiti and hence the ideal person to help with my study. People did not know her and were less likely to be biased in their responses. Saloni's experience of studying how the media reported on common leopards would also help. And this study would give Saloni the necessary experience of working in the landscape before she embarked on her own PhD. Once Saloni took over the interview surveys in Spiti, I decided to focus my attention on another study site, Ladakh, to consider it for my dumpling study.

Ladakh is popularly called Little Tibet. It is the westernmost extension of the Tibetan plateau and at the crossroads of Central Asia. Many great journeys have passed through here. Further west of Ladakh across the greater Himalayan range is Kashmir. To the north, across the Karakoram and Pamir mountains, are the historical cities of Kashgar, Dushanbe, and Tashkent. To the south across the Zanskar and then the Himalayan ranges lie the fertile plains of the Ganga and the Indus river valleys.

As the British Empire expanded in India, and Tsarist Russia expanded in Central Asia, the two giants were set to collide in the mountains of Ladakh. Thus began what is known as the Great Game between these two expansionist regimes. The first empire to understand these mountains would enjoy a huge advantage over the other. Thus explorers, mapmakers and spies like Sir Francis Younghusband and Pundit Nain Singh Rawat went on expeditions in the mountains around Leh, the headquarters of Ladakh. On his famous expedition across Asia in 1886 from Beijing through the Gobi Desert, the mountain cities of Central Asia and then across the Mustagh Pass in the Karakoram mountains, Francis Younghusband passed through Ladakh to Srinagar. This is considered one of the greatest crossings of the Central Asian mountains. Explorers like Nain Singh Rawat were sneaking into Tibet and other Central Asian kingdoms, dressed as monks, pilgrims or traders while measuring their path by counting their steps and using trigonometric instruments to map this unknown land. Even their scientific instruments were disguised as religious paraphernalia like prayer wheels.

Ladakh is a place of contradictions. Cooped up between some of the greatest mountains in the world, it has also been the intersection of great civilizations, cultures and empires. Although difficult to reach, great caravans passed through laden with riches such as cashmere, wool and gold from mountain kingdoms. Three nuclear power states, India, China and Pakistan, are today in conflict over these lands, deploying some of the largest armies of the twenty-first century to defend the territory.

I took a flight from Delhi to Leh. It would be my first time flying over the Himalaya and I booked a window seat with great anticipation. As the flight took off from Delhi and gained altitude, we left behind a vast layer of dust and pollution. Soon I could see small white mountain-tops, emerging like crystals out of a layer of grey clouds in the distance. I was looking out for peaks that I could recognize. We flew over Spiti and I could make out some of the familiar villages like Kibber and Tashigang.

Across the great Himalaya, with its high peaks and glaciers, the plane flew over the vast Changthang plateau. This region is made up of vast rolling hills with scattered mountain ranges and rounded summits. From our altitude, it seemed that the region was covered in a layer of dust. There was not a speck of snow anywhere. Soon the pilot announced that the flight would descend into Leh.

Leh is located on the right bank of the Indus River which originates near Manasarovar Lake in Tibet and flows west. By the time it reaches Leh a couple of hundred kilometres later, this large river shapes the landscape. The city sits on the warm south-facing terraced slopes and the Indus laps the agricultural fields at the feet of the city. Across the river, the Stok Kangri peak stands tall at over 6,000 metres. Its triangular summit looks deceptively within reach of Leh.

Pilots must pull a complex manoeuvre to land the Airbus A320-200 on the airstrip in the narrow valley at Leh. The aircraft did a perfect circle to lose some altitude, then banked hard on the right, flying dangerously close to the mountains just north and above Leh before making a full turn to descend into the valley coming up along the Indus river. Palpable excitement filled the aircraft. Nervous fliers kept looking ahead, those more adventurous were pointing out little houses, monasteries and mountains through the windows, and I was experiencing motion sickness. I could not wait to get back onto stable ground. As the flight approached the runway for landing, the aircraft glided smoothly between a high mountain on one side and a monastery on a hill on the other. The monastery was far above as if suspended in mid-air, even while the aircraft was still flying.

We landed and the pilot announced it was a crisp −11° C outside. The small two-room airport was full of noticeboards about the dangers of high-altitude sickness and advised travellers to rest for the first twenty-four hours. Outside the airport, I was greeted warmly by Karma Sonam. Karma had joined NCF many years previously, in 2005. He was at least twenty years older than me, and I always addressed him as Karma-ji. The years of working outdoors in the harsh sun of the high elevations had left his face with a deep tan and wrinkles radiating from the edge of his eyes and the corner of his mouth. He had a big smile, and he welcomed me by placing an *aashi* (a white satin cloth) over my shoulders as a mark of respect.

I had come to Ladakh to consider it for my dumpling study on snow leopard diets and livestock predation. Karma and I planned to trek from his village of Rumtse through an area called the Tsaba Valley, across a high pass and descend back towards the Indus River.

The Tsaba Valley was the wintering pasture for the fourteen semi-nomadic herders from Rumtse and the surrounding villages. They leave this place ungrazed through spring, summer and autumn so there is enough grass for their goats in winter. The region receives only light snow and remains accessible on foot throughout the year. I had arrived here in the middle of November, and after two days of acclimatisation we drove to Rumtse in Karma's car. Rumtse is at an elevation of 4,200 metres, about 700 metres higher than Leh, and I had to spend another two nights acclimatizing. We were planning on crossing a pass at 5,500 metres made all the more difficult by November's early winter cold.

The first day of the trek was exciting. We walked along a spur replete with snow leopard signs and we could spot four different herds of blue sheep on either slope. I was happy to see that the blue sheep were not afraid of us, which meant that people were no trouble for them. We were about to reach our camping spot when Karma sat on his haunches and looked intently through his binoculars. I sat down next to him and looked in the same direction. The slope in front of us was gentle and had a faint wash of fresh snow. I could see the hoof marks of a herd of herbivores. Karma-ji said, 'Those tracks were not made by goats. Goats move parallel to each other and leave a uniform pattern of tracks. This is too erratic.'

'Blue sheep?' I asked.

'I have never seen blue sheep on that slope,' he replied softly.

We kept looking through our binoculars and we both noticed a movement almost at the same time. At first, it looked like a single animal standing with its head lowered, but then we noticed three or four others sitting next to it, and as our search vision got better we noticed nearly twenty large animals sitting and chewing cud.

The animals were argali sheep. The argali is the largest species of wild sheep in the world. Their long, supple legs help them run over the rolling hills in the thin air of the high mountains, and the males stand tall at over 6 feet. When Marco Polo saw one during his travels in Central Asia, he described the animal in great detail in his diary, and today the sub-species of argali found in Tajikistan and parts of Kyrgyzstan are called Marco Polo sheep. Here in Ladakh, the males grow thick horns that curl behind their ears and then point out in front from below their jaw. These are the biggest sets of horns of any species of wild sheep and goats in the world. Argali visit the Tsaba Valley only during the winter. They cross the mountain range in the south before spring to give birth to their young on the banks of Tso (lake) Kar.

It was getting late and we had another hour of walking before we would reach our campsite, so we left the argali sheep and continued on our trail. Karma pointed out the camping site in the distance and we picked up our pace. Again, in the light snow at our feet, we could see a set of hoof marks running across the trail. At first we did not pay much attention, but then we spotted a solitary raven sitting on a rock in the direction of the marks. Curious, we walked towards the raven in a small boulder field. In between a couple of boulders lay a dead argali. The flesh was gone, leaving behind a set of bones, dry hide and a pair of majestic horns. The entrails had been pulled and strung out in a long line by foxes. The hide was still soft in some parts, suggesting it was not very old. What animal could have consumed a whole male argali in such a short time?

'Wolves?' I asked, guessing that a pack of wolves would finish an argali in a single meal.

'No, they tend to leave a lot of mess behind. There would be pawprints all over the place. They fight each other while eating and scatter snow and gravel. Believe it or not, I think this is the work of a snow leopard,' Karma said.

We looked around for pug marks but couldn't find any. The snow leopard would have moved over the large boulder without leaving any signs. Over the years I have learned that snow leopards avoid stepping onto patches of snow if they can. Sometimes they go to a lot of trouble to avoid deep wet snow. When forced to walk over powdery snow, their long tails leave a snaking impression between their pug marks, but their magic trick is to walk over hard snow where their wide paws don't leave any marks. I have seen snow leopards and then searched in the place where they stood minutes before and not found any pug marks. This trick adds to their legendary ghostly status.

The sun was behind the ridge and it was getting cold. We decided to keep walking. Walking would keep us warm.

'I think Tashi is already camping here,' Karma said when he saw an old yak-hair tent at the camping spot. A thin wisp of white smoke was coming out of the small chimney of the tent. The possibility of warm chai after a long day of walking spurred me on.

Tashi must have heard our footsteps for he came out of his tent and greeted us with a cheerful *jullay*.

Tashi was a small, thin man with bow legs. He wore heavy jackets covered in a layer of dust mixed with sweat, forming a crust. I guessed him to be younger than he seemed. He was clearly delighted to see Karma. They were busy in an animated conversation with lots of laughter but I could not follow the dialect of Ladakhi they spoke. From his gestures, I could make out that Karma had introduced me and I said a polite *jullay* with my palms joined and head slightly bowed. I was getting cold and I could not wait to be invited inside for a chai. Once inside, the animated conversations continued and I was served a steady supply of both sweet and salted butter tea.

Unlike the migratory Changpa herders of western Tibet, who travel with their entire family, the herders of Tsaba travel alone with their livestock while their family members live in permanent villages like Rumtse. Their pastoral system is transitioning from a migratory to sedentary way of herding. Herders like Tashi were lonely and bored. His traditional yak-hair tent was warm and kept all the light and wind outside. It was small, almost like a three- or four-man alpine dome tent with a hole in the middle to fix a small metal chimney. A small place of worship with a picture of the Dalai Lama was reserved at the back of the tent.

After the first round of chai, Karma told me that Tashi was suggesting we stay with him in his tent instead of pitching our tents. He was slightly hesitant in suggesting this, perhaps because he thought I needed my privacy, but I was delighted at the idea of sleeping in a warm tent with a small fire. My tent, with its thin synthetic fabric, was designed to be light rather than warm. It would keep the wind out, but nothing more. There was also no chance of keeping a fire going inside it.

Then Karma mentioned that the dead argali we saw on our way was indeed killed by a snow leopard about a week ago. Tashi had bumped into the snow leopard the day it made the kill. Tashi had been happy that the snow leopard had a large prey for itself and wouldn't be bothering his goats for a few days at least, but last night Tashi had seen the eye shine of a snow leopard near his flock again.

Tashi's flock was corralled in an open pen right behind his tent. I had missed it when we arrived because I was so cold and I needed a chai. After a simple dinner of *tsampa* (paste of roasted barley with butter and water), I got ready for bed. Karma-ji and I rolled out our sleeping bags while Tashi lay down on his heavy bedroll.

I dozed off while Karma and Tashi were still talking. Karma was Tashi's window to the world. He brought news of marriages, births, deaths, new houses and fights and feuds in the surrounding villages.

I woke up in the middle of the night to the sound of a sharp whistle. I switched on my headlamp and saw that Tashi was not in the tent. Karma and I looked at each other and said 'snow leopard'. It took me a long time to dress for the –20° C temperatures outside.

By the time I was out of the tent, Tashi was already on his way back. He had heard the dogs bark and had gone to check on his goats. He shouted a few times to scare away any snow leopard or wolf hiding nearby. He was not even carrying a staff or a stick, let alone a weapon to scare the carnivores. Just his voice and the stones that he would throw at them with a sling made from yak hair.

I removed my feather jacket and got back into the sleeping bag. I must have dozed off for a few minutes when we were woken again by the loud barking of the dogs. We rushed outside. There was some commotion in the goat pen. We looked around for eye shine but could not spot anything. I was cold as soon as the excitement ebbed. Tashi said we should go back to the tent while he stayed out for a little longer, but Karma decided to stay back with our host for company. I was torn between going back to the warm sleeping bag and staying with Tashi and Karma.

It was exciting to think that I might get to see a snow leopard, but it was late and cold and the goats and the dogs had kicked up a cloud of dust which scattered the light of my headlamp and blinded me. We waited in the dark and slowly the dust settled. We switched off our lights to let our eyes get used to the starlight. A dark outline of the mountains around us emerged against the lit sky with the twinkling of millions of stars. Suddenly the flock of goats moved as if pushed by an unseen hand. It startled me but Karma and Tashi were alert and peering into the darkness around us. I could see the fear on the faces of the goats; frightened, they huddled as if trying to get away from the darkness.

After a long bout of silence, I wanted to go back to my sleeping bag in the warm tent. I walked around with Karma-ji and Tashi for a little while and then they decided to return, mostly because they felt sorry for me. Tashi knew he would have to return to his goats but he came along. Back in the tent, we did not change. Tashi sat on his bedroll and was fidgeting around with something. Karma lay down on top of his sleeping bag. I got into mine but did not remove my winter clothes. In another fifteen minutes, we were back outside: another round of barking by the dogs. We did this all night, finally getting some sleep after 4 a.m.

We did not see the snow leopard that night but the goats and the dogs were restless, feeling its presence around them. That kept Tashi and Karma anxious and all of us sleepless.

Tashi woke me up some time around 8 a.m. He had a hot cup of chai for me. The tent was very warm. Karma was already sipping his chai, and he moved a bag of *tsampa* towards me. I added some to my chai after taking a few sips, turning it into my breakfast. There was a sense of urgency to Tashi and Karma's movements. Soon Tashi headed out with his goats and a few dogs. Karma and I packed our backpacks and headed up the trail towards the pass. Today we would have to climb from 4,600 metres to the pass at 5,200 metres and climb down another thousand metres to our destination. We had walked all day yesterday and stayed up for much of the night. I was worried about my ability to make it across the pass. I braced myself for a hard day ahead and started putting one foot in front of the other.

Saloni and I caught up in Bangalore. She was pleased with her fieldwork experience in Spiti and wanted to build on this effort for her PhD. Between us we had talked to over four hundred people in Spiti to understand their attitudes towards the snow leopard and the wolf. My eyes lit up when Saloni showed me the Excel sheet that displayed the numerical interpretation of the information.

Two things stood out. This was not a simple story of snow leopards killing livestock and herders hating them for it. Neither was it a story of compassionate Buddhist people tolerant of anything and everything wildlife does to their livelihoods.

People's attitudes towards snow leopards and wolves spanned the spectrum of very positive to very negative. What surprised us was that the most common response to the question 'What should be done when a snow leopard kills livestock?' was 'They need to eat too!' A statement which herders often qualified, saying as long as snow leopards only occasionally killed the weak and old, or as long as they did not enter the corral and kill livestock en masse, or as long as the herders were compensated for the kill by NGOs or the government.

I started exploring the data further. The numbers – in all their abstraction – were telling a bigger story than the impressions that Saloni and I had gathered through anecdotes. Herders who lost more livestock to snow leopards did not think worse of the snow leopards. This did not make sense. But looking again I realised that herders from villages which lost more livestock overall to snow leopards thought more negatively about them. Our lived experiences are not ours alone. We make sense of the world using the experiences of people around us and close to us.

We also learnt that people had better things to say about the snow leopard than the wolf, even when snow leopards killed more livestock than wolves. This was surprising, but made sense. I reviewed literature from around the world and realised that wolves are consistently portrayed as the villains in the stories and folklore we tell. This may have to do with wolves being less secretive than large cats; their howling behaviour makes them more conspicuous and, as pack hunters, their method of killing and eating their prey seems more cruel. Our data showed that wolves were taking the blame for snow leopards in some places.

But I was caught off-guard by the finding that women's attitudes towards the snow leopard were worse than men's. The pattern remained consistent even when I separated the data I collected and the data Saloni collected. The gender of the person collecting the data did not affect the patterns. We had ensured that at least a third of all the people we spoke to in a village were women. In the end, 43 per cent of the survey participants were women, so the patterns were not due to a small sample size. Suddenly I was reminded of the words of the woman I met in the basket suspended many hundred feet in the air: 'Why don't you marry one of your snow leopards?' She had told me how she had spent sleepless nights guarding her animals and yet there was no milk in the house for her child. And the first person I spoke to during this study had told me the story of the yak she had raised from birth, only to see it taken by a snow leopard. While NGOs like NCF and the government had put in place mechanisms to financially compensate for these losses, the emotional burden of living with snow leopards was being borne by women.

Social scientists working in other parts of the Himalaya were calling it the hidden cost of living with wildlife and finding that women face it disproportionately more. Looking deeper into our data, we found that even in villages where we had conservation programmes, women's attitudes were more negative than men's towards the snow leopard and wolves, although women and men in these villages had more positive attitudes than people from other villages where there were no conservation projects. So, NGO-led conservation projects were effective, but not effective enough to help men and women equally. The issue was multifaceted. Growing up, girl children in rural societies have poorer access to good education and nutrition. Women were taken out of school earlier and had less access to the household cash income. Combined with this, when a carnivore was around, women bore the social and psychological costs of staying up late, worrying about the safety of everyone in the house and being responsible for feeding their family the next day.

This study caused me to reflect on myself as a scientist and conservationist. The outcome of our good efforts is often subject to our biases, which are reflected in whom we interact with in society. While I identified myself as an egalitarian scientist and conservationist, my actions revealed that I had been shaped by a complex interaction with the local society. I had found comfort in thinking that monetary compensation at the household level solves the problem of livestock predation by snow leopards, without realising that it was mainly solving the problem for the men of the house.

The findings of this study came around a time when another prospective doctoral student on our team, Ranjini Murali, was getting ready to bring more women into our conservation initiatives in Spiti. One of our senior colleagues in Mongolia, Bayarjargal Agvaantseren or Bayara, had pioneered a women-led business called Snow Leopard Enterprises to engage women in snow leopard conservation and help them raise cash income for themselves. This initiative had been running successfully for over ten years.

In India, Ranjini Murali initiated discussions with the women of Spiti about the possibility of a similar women-led enterprise that could produce and sell local handicrafts to raise awareness about snow leopard conservation and to make money for themselves. The overarching goal of the project was to involve women in the local conservation discourse. Such an enterprise could be an avenue for women to express their concerns during our discussions about conservation. These conversations led to the formation of Shen – *shen* meaning snow leopard in the local language – an initiative of the Snow Leopard Enterprises. Shen organised training for women not only in creating handicrafts but also in financial literacy, running small businesses and arranging exposure tours for them to Delhi and Bangalore. For some of the participants, this was the first time they had stepped outside their village.

After initiating the project, Ranjini started her PhD with a study on the ecosystem services that people derive from snow leopard habitats in the high mountains of Asia, and handed Shen to the able leadership of Radhika Timbadia. Radhika developed Shen into a platform for women to organize themselves into self-help groups. One of the first groups organized under this initiative proudly called themselves 'Golden Eagles'.

Within a year, two groups of ten women had made over $3,000 through the Shen programme and they were also participating in on-the-ground conservation activities. I still get goosebumps remembering an incident when five women from the Ama Chokspa group in Kibber village went to stop twenty-odd men, who had come from outside Spiti to work on a road construction project, from killing a blue sheep. The five women stood their ground in the altercation and threatened the men with jail if they ever did something like that again. Over the next eight years, Shen grew to include over a hundred women from seven different villages, and in the next decade these women from the remote villages of Spiti would go on to play a crucial role in estimating the population of snow leopards in the region.

Reflecting on the findings of our study, it was easy to think of the more negative attitudes of women to the wild animals that encroach on their lives as a problem only found in rural places in remote regions of the world. Researching this phenomenon further I was surprised to find that women's negative attitudes to dangerous wild animals were far more widespread. Conservationists and psychologists from places such as urban centres in Norway were reporting that women were less likely to support the conservation of wolves.

When I had begun my career in wildlife research and conservation, I thought that I was leaving the world of social problems behind me, and entering a place of unadulterated nature. What I was beginning to understand was that wildlife research and conservation does not happen in isolation, away from the rest of humanity, but is intricately bound up in the lives of the people who live in these remote places.

# 5

# Aztai

A loud alarm, like the siren of an ambulance, woke us in the middle of the night. The three of us were up in a split second. Our *ger* – a traditional Mongolian portable round tent made of felt and canvas – was still warm. Örjan switched off the alarm and started getting ready. I went from a state of fast asleep to palpably excited within a couple of seconds. The triggered alarm told us that an animal, hopefully a snow leopard, had been captured in one of Örjan's traps. I knew the drill. I threw on a heavy jacket, picked up the radio antenna and headed up the hill behind the camp. It was a little after midnight and tonight it was my turn to go to the receiving station to figure out exactly which trap had been activated.

In 2011 I was in the Gobi Desert of Mongolia, at the long-term research station jointly run by the Seattle-based Snow Leopard Trust and the Snow Leopard Conservation Foundation, a Mongolian NGO. The station had two camps. A base camp in the middle of the Tost-Tosonbumba mountains had five *gers* and a large shipping container. The *gers* were powered by solar panels and set up to act as workspaces, dormitories for researchers and students, and small laboratories. The container was used for storing ration and research equipment. Midjii and his wife Oyuna lived at the camp with their daughter and worked as base camp caretakers. Oyuna was the daughter of a herder who had a camp nearby and Midjii had retired from the army after the fall of communism. There was also a second field camp with two *gers* that was moved around closer to the areas where the team was doing fieldwork. The base camp was larger and better stocked and useful for doing field trainings for university students and rangers, while the field camp was nimble and used for data collection in the more remote parts of the desert. I had arrived at the field camp only a few days before after a two-day drive from Ulaanbaatar, the capital city, to work with Örjan Johansson, a PhD student from Sweden who had mastered the technique of safely catching wild snow leopards to fit with GPS collars in order to study how they used their habitats, how often they hunt and where females make a den when they have cubs. Örjan had set up the field camp in the south-east corner of the mountains, about an hour and half's drive from the base camp. Assisting Örjan was a veterinarian from Australia working as a volunteer. My goal was to learn from Örjan with the hope that these skills could be used in my work in India. More importantly, I was to set up a long-term study here to monitor the populations of herbivores using the Double Observer method; this would be one of the seven sites I included in my PhD research.

I had walked a couple hundred metres up the hill towards the radio receiving station when a pair of eyes lit up in the glare of my headlamp. I froze. The eyes were barely 20 metres away. Their front-facing position indicated that the animal was a carnivore. Slowly, another pair of eyes shone out behind the first one, and then another. A pack of wolves this close in the dead of the night could be dangerous. But they were snow leopards, a mother with two cubs. I was expecting her to retreat but she waited for her cubs to catch up and then headed at an angle right past me, almost going towards our camp. She would skirt around our camp and climb the ridge on the other side. She seemed to be in a hurry.

I was also in a hurry. Whichever animal had triggered our trap had been stuck in it for about five minutes. The longer it stayed in the trap, the greater the chances that it would hurt itself in its attempts to escape. We had to be quick. I watched the snow leopards disappear into the darkness and then kept climbing, breathless. I checked the receiver systems and found that trap number seven was triggered. It was on the other side of the receiver – the direction from where the female and the two cubs had come. I rushed back to the *ger*, running downhill in the dark. Örjan and the vet were ready.

'It's trap number seven. It is on the other side of the ridge from the receiver,' I said. Örjan nodded in agreement but double-checked a laminated map that he always carried with him. His methodical approach made him very good at safely catching wild snow leopards.

'There's something else. I saw Khasha and her cubs. They came from the direction of the trap and headed to the ridge in front of our *ger*.' The vet looked excited. The chance of seeing wild snow leopards can create a visceral reaction in the body. But Örjan's face fell. 'Ah, it is possible that Khasha or her cubs triggered the trap but were not caught in it. They must be rushing because the snap of the foot snare startled them.' In his desert camouflage clothing and with the dart gun, Örjan looked like a soldier from a special ops unit. Örjan and the vet rode in one ATV and I took the other with a rucksack on my back. There was not enough time to carry all our gear and walk over the ridge to trap number seven, so we were going to ride on the ATVs around the ridge and get there faster.

Our path followed a dry, sandy river bed. There were no roads here – this was like riding the Dakar Rally at night. We were within a few hundred metres of the trap within twenty minutes of the ringing of the alarm. We left our vehicles and started walking on foot. Örjan with the dart gun in the front, followed by the vet, then me. We tried to stay as quiet as possible. As we got close, I could see a pair of eyes shining in Örjan's headlamp. A snow leopard was caught, its forefoot held in the snare. It had seen us but it was surprisingly calm.

With the smoothness and finesse of an artist, Örjan kneeled, calibrated the pressure in the dart gun, checked the dart and the drug dose in it, took aim, and shot the dart in the right thigh of the snow leopard. Surprisingly, the snow leopard barely reacted to a dart hitting his thigh, sanguine to his fate like a grumpy kid who has been forced to take a vaccine shot against his will. We stayed a few seconds to ensure that the dart was lodged correctly and the medicine was injected properly. The snow leopard stayed still like a ballet dancer in the spotlight – muscles taut, ears pricked, tail poised, the soft rosette-pocked fur gently draped over its shoulders. We turned around and walked a little distance to let the medicine do its job. A stressed animal could try to fight the drug.

While we were waiting, Örjan spoke first. 'I think this is Aztai, the large dominant male from this region. This is the seventh time I have caught him. You saw how calm he was. He knows the drill.' I felt embarrassed that I had not noticed the collar on him when he was darted. We pieced together what must have happened. Aztai must have been following Khasha and her cubs, who are almost grown up. Somehow, the mother and cubs walked past the trap, which was set at a scent-marking spot. Khasha was not ready to mate so she did not mark at the site and escaped the traps. But Aztai would have gone to check the scent-marking site and got caught. Khasha and her cubs were rushing not because of the trap but because they wanted to get away from Aztai.

After about fifteen minutes, it was time to check on Aztai. When we reached the spot, he was fast asleep and we sprang into action. We had about thirty to forty-five minutes before Aztai would start waking up. We found a flat spot a little distance away. This was my first capture operation and I had gone over the drill in my head many times, but I was nervous. Örjan asked me to carry Aztai to the mat that was laid over soft ground. I lifted him with both arms and held him in front of me. Before I started walking Örjan reminded me that that another person needed to hold the tail. Although I was holding Aztai at my chest height, his tail reached the ground at my feet and I could have easily tripped on it and hurt us both. Örjan held the soft bushy tail like a bridesmaid holding the train of a bride's dress and walked alongside me till we reached the mat and carefully placed Aztai down.

The first thing to do was to check the pit tag, a small device half the size of a shirt button, that is inserted under the skin at the back of the neck and which has a unique number. Örjan scanned the snow leopard's neck with a small device that looked like a metal detector and when the number flashed on the screen, he confirmed that this was Aztai. While Örjan was busy with this, the vet set up the equipment needed to monitor Aztai's body temperature, his breathing and other key vitals. The spot on the sands looked like a small operating theatre within a matter of minutes. Aztai had been carrying his old satellite collar for many months and Örjan reasoned that the batteries would run out soon so we should change the collar now. The vet collected blood samples for genetic and molecular analysis. I was handing over all the necessary equipment as needed, like a nurse in the operating theatre. Finally, we weighed him. Örjan guessed his weight correctly as 40 kilograms.

We were finished sooner than expected and it was time to wake him up. Örjan injected him with the antidote and we packed and moved the equipment and ourselves a hundred metres away near to the parked bikes. Within minutes of receiving the antidote, Aztai opened his eyes. He would need some time to be fully awake. The anaesthesia would wear off from the front of his body to the back, so he would move his head first, the neck, then his front legs. If disturbed during this time, he would try to get away, dragging his hind legs and tail using his strong front paws. We needed him relaxed and lying down. Aztai kept lifting his head and looking around but did not try to stand up. This was a good sign.

Since the alarm, everything had happened fast but already it was a good hour and a half since we had left the comfort of the *ger*. Now I remembered how cold it was. Late in October and close to 2 a.m., the desert was freezing. Finally, Aztai got up. We held our breath to see how he walked. He stepped forward confidently but his hind legs looked weak, his gait like a drunken reveller. With every step, his hind legs gained more control. It was like a scene in a sports movie where an out-of-shape former sports star transforms into a strong athlete with each step of their morning run.

When we reached our *ger*, I asked Örjan how long the drugs would affect Aztai. He said that Aztai would have a hangover for a day and then he should be fine. 'How long before he can hunt?' I asked. Örjan smiled at my concern and went on to narrate a story. The first time he had caught and collared Aztai, he noticed from the tracking device that Aztai did not move from the collaring location for two days. He returned to the spot and found that Aztai had killed a large male argali and his belly was stretched from all the meat he had eaten. Aztai did not move for another week. So Örjan went to check again and saw that he had killed another huge argali at the same spot. 'Aztai seemed angry and was venting it on the poor argali,' Örjan joked.

The following week, we caught Khasha and replaced her collar. We caught Aztai once again and released him from the trap and immediately gave him the antidote. He seemed the most gullible to Örjan's tricks. Örjan was a master of his science. By this time, he had captured and collared more snow leopards than all the other biologists in the world put together. What was more impressive was that no injury or harm had befallen to a single snow leopard or to a team member. Every time a human patient is put under anaesthesia, the hospital has them sign a liability waiver as there is always a small risk that things will go wrong. This risk is amplified when we work in the wilderness with dart guns for injections, makeshift tables in the sand and flashlights for illumination. Working with Örjan, I learned that he was working to minimize the risk at every step of the process. When he chose a site to deploy the trap, he would assess it for the leopards' safety. He deployed foot snares that would spring out from the ground when the leopard stepped on the right spot. Örjan would spend hours thinking about the possible ways a leopard would approach a snare and the things that could go wrong. Together with his engineer brother, he had designed a radio system that alerted him within minutes of the snare being triggered. The less time an animal spent in a snare the lower the risk of harm through trying to get free. Örjan also worked with the latest set of drugs, known to be the safest through multiple clinical and field trials. He treated his equipment with the same amount of care. He was constantly fixing things and improving the camp. He believed that if one thing went wrong it could set off a cascade of mistakes. Far out in the middle of the Gobi Desert, we could not afford such a thing happening. And yet, a few weeks into my visit, things got a bit derailed.

The collars that Örjan put on the snow leopards transmitted their GPS location to a satellite rotating around the Earth in space. The satellite sent the data to the Snow Leopard Trust's head office in Seattle, where it would be put in an email and reported to the base camp through a satellite-phone-based email system which transmitted a few kilobytes of data every minute. Örjan would plot these locations on a map and every time the locations clustered in one spot, we would know that the leopard had spent a long time there, most likely because it had made a kill and was eating it. We would drive to these spots a few days later and from the remains of the carcass discover what the snow leopard had killed, record the characteristics of the location and learn everything we could. We would visit the feeding sites during the middle of the day when the snow leopards were least active and the chances of one of them getting into a snare back at camp – from which we were now far away – were the lowest.

We were on our way to a cluster of locations far away from the camp. Örjan rode his 250cc motorbike with the vet as pillion. I was on the other bike. There were only a few rudimentary roads in the desert that the local herders used with their large Russian vans or Chinese motorbikes, nothing more than two lines in the dirt. Every time we hit a sand patch, those lines disappeared. Riding in the dirt was easy and fun; you had good control of the bike and the wheels had a lot of purchase, but riding in the sand was another ball game. If you were too slow, the wheel would sink in the sand and either stall the bike, or worse, the front wheel would turn and topple the rider over. You therefore had to ride fast enough that the wheel wouldn't sink but only as fast as you could control. Making a turn on these sandy stretches was very risky. Örjan would say that we had to imagine a Rammstein song in our head when riding the bike through the sandy section.

On these long forays away from the camp, we would take the roads that brought us closest to the spot with the cluster of locations, then use a handheld GPS to ride as close as possible over rocks and dirt and dunes. The last stretch we usually made on foot.

We were around 20 kilometres away from the camp on a dirt road when Örjan lost control of his bike on a sandy stretch. In a blink of the eye, he and the vet were on the ground. We had taken such falls before so I did not think much of it as I approached them. The vet was already on her feet but Örjan was clutching his shoulder. I picked up his bike and the equipment that was scattered around. The maps, the GPS, a piece of plastic broken off the bike. The vet was checking Örjan.

We huddled together to assess our situation. Örjan had either broken his shoulder or dislocated it badly. We were far from the camp. We made a quick decision. I was to ride back alone to the camp and drive back in an ATV to return us all to the camp. We did not want to risk another accident on the bike with a pillion rider. ATVs have four wheels and are safer than bikes but they are slower to ride and not great for travelling far away from the camp, as they also require more fuel which came from a mining town over 50 kilometres away. I made the trip and was back in a little over an hour and found Örjan and the vet walking along the dirt track. For a few seconds I thought he might be better but I was wrong. Although he was walking his arm was in a makeshift sling, unable to bear its own weight.

It was night by the time we arrived in the camp. On the return journey we had tripped the snares that were within a short driving range of the camp. We did not want a snow leopard in the traps while Örjan had an arm in a sling. A couple of traps remained primed further out but we decided we didn't want another accident and it was now dark.

Early in the morning, the alarm went off. This time we were shocked awake. The alarm should not have sounded: we had tripped most of the traps. I went to the receiving station on the hill to see which trap had been triggered. It was trap number two: one of the farthest from the camp; one of the two that were still active. By the time I reached the *ger*, Örjan and the vet had a plan. Örjan was confident that he could dart the leopard safely. Dawn was rising so the visibility would help. He had tried a few shots on an old blanket hung outside the camp and succeeded in hitting the bullseye every time. The vet was confident of handling everything from that point on. By now I was well trained to handle the satellite collar which would be placed on the leopard. The vet, with Örjan as pillion, rode one ATV and I took a motorbike.

When we reached the trap, we realised it was a new snow leopard of a smaller size. We suspected it was one of Khasha's cubs. Örjan perfectly shot the dart in the middle of the leopard's thigh. Everything went as planned. We gave him a pit tag, for the time being calling him M9 (male number nine). The vet had managed his body temperature and breathing just right. He was young, as we had suspected, weighing close to 30 kilograms. His body condition was perfect. The vet and I put a new satellite collar on him. We had to get the circumference right so that there was enough room for his neck to grow without any discomfort while making sure that the collar did not slip over his head.

It was time to wake up M9. We moved to a place where we could watch him but not disturb him with our presence. M9 lifted his head but stayed on the mat for a long time, resting. The sun had risen and a golden light washed the stony hills around us. Just as the sun crested the ridge, M9 stood up and walked along a spur leading up the ridge on the east. The early morning sun hit his face and he looked like a prince about to be crowned king of the mountains.

Thanks to the satellite collars, Örjan wrote a series of influential research papers. His data showed that male snow leopards were ranging over 200 square kilometres while females ranged a little over half of that. But shockingly, he also showed that 40 per cent of all the protected areas in the high mountains of Asia were too small to accommodate the home range of a single male. These protected areas were not designed to protect a large carnivore. Many were made to protect small valleys with high biodiversity and natural beauty. Moreover, only one in ten protected areas were large enough to have a small population of fifteen snow leopards. Although human population is low, the pressure on land for livestock grazing, mining, military activity and, increasingly, for renewable energy is very high – limiting the size of the protected areas.

In the popular imagination, snow leopards live in remote faraway mountains undisturbed by humans, but in reality there is no place for snow leopards to live in complete peace. Snow leopards need large spaces with sufficient wild herbivore prey to eat to survive. To successfully breed and raise cubs, a female needs territory with high densities of prey where she can hunt at night and be back with her cubs during the day to protect them from harm and suckle them until they are old enough to walk with her to her kill. As the cubs grow, the need to hunt becomes more frequent. A fast-growing one-year-old cub could eat almost as much as an adult. A mother snow leopard with two or three cubs will need to hunt twice or even thrice a week. The satellite collar data showed that snow leopards were killing one ibex or argali or domestic goat every eight days, about fifty kills a year. The territory of each snow leopard had to be large enough with enough prey to sustain extraction of fifty animals each year and not deplete. And if they ever killed a goat or a horse belonging to a herder, then they could face serious persecution. One of the collared snow leopards was shot for this reason.

The area over which a snow leopard ranged, how much prey they needed, how often they hunted, how often they bred and how many cubs they had each time were some of the important details we needed in order to understand how many snow leopards there were in the world and how they were faring.

Months later, Örjan emailed me to say that M9 had dispersed. He walked nearly 50 kilometres over flat desert land to reach a distant mountain range in the north – a range so far away that you could barely see it from the tallest peak on a clear morning. But M9 had seen it and made a beeline for it. He would settle down there and establish a home, just as Aztai had done here.

# 6

# Cashmere

The safe collaring of M9 was as much a relief as a joy. Despite the accident, we had been able to perform a complex operation prioritising the safety of the leopard. After the collaring, we dismantled the remaining trap and brought all the gear back to the camp.

Örjan needed treatment and the nearest MRI machine was in Beijing (although Ulaanbaatar Hospital had one we could not get reliable information on whether it was in working condition). Örjan decided that he might as well fly back home to Sweden and begin the treatment from the comfort of his house. The decision was made: Örjan was to be evacuated. The vet would leave with Örjan. I had to decide whether I wanted to stay in the field camp alone and continue my work.

Exploring the Gobi Desert alone while on a scientific expedition was very appealing. This was my moment to walk in the footsteps of legendary adventurers like Messner and Younghusband. I would not be looking for trouble in the name of adventure, but doing an important task. I was not going to pass on this opportunity, and I decided to remain in the Gobi. It was a risky decision. If I had an accident while I was in the mountains, I would not be able to reach the outside world. I would have to make it back to the field camp to the little satellite-phone-based email system. Midjii and Oyuna would be at the base camp 14 kilometres away. Midjii would visit me every few days to check if I was okay.

Orjan and the vet departed the next morning in a large grey

Russian van with Midjii. The drone of the engine hung in the air for a long time until silence took over. I knew what I had signed up for, but only now did I begin to understand how lonely it would be. The trick was to stay busy. I started reorganizing the camp and thinking about what I needed to do over the coming days and weeks. But my thoughts kept going back to M9. He was about the same age as Shadow when I had last seen him.

My nearest neighbour was a kind herder named Baatar who had his camp some ten kilometres west from Örjan's *ger*, which I now made my home. Baatar was Midjii's father-in-law, and we often shared pleasantries when Midjii came to check on me. We would share some tea and meat in silence before going our separate ways, him after his goats, and me after my ibex. The next nearest neighbour was a similar distance in the east. We hardly ever saw each other.

The Gobi Desert is the farthest place from any sea on Earth. So little moisture reaches the desert from the coast that it is one of the driest areas in the world. Harsh winds have weathered the Altai mountains into a series of hillocks and boulders strewn between sand dunes and oases. Siberian winds from the north bring some winter snow and an occasional rain in spring, which support a unique assemblage of plants and grasses. This combination of a rugged mountain range and sparse vegetation is ideal for wild goats and sheep like the ibex and the argali and their predator, the snow leopard. But the most unexpected residents of this desert are domestic goats. These goats outnumber the ibex and argali by a hundred to one. Semi-nomadic pastoralists raise these hardy beasts in this windy and cold place so that they grow the warmest cashmere undercoats.

Around this time, my wife Bhagya and I were still in our courting phase. She was interning with a team of ornithologists surveying for hornbills in the rainforests along the border between India and Burma. I was fortunate to have the satellite phone in my *ger*. I was not allowed to make calls except for an emergency, but I could send and receive short emails. I would wait every day to receive an email from Bhagya before drafting my careful reply.

Then I would send a daily email to my office in Bangalore assuring them that I was okay.

I had just begun my PhD research when my mother introduced me to Bhagya in 2010. She was then in the fourth year of her bachelor's degree in fisheries science. We met twice before we were engaged. Within months, I left for my first field trip to Mongolia.

Bhagya and I shared an enthusiasm for the outdoors and sports. One evening, Bhagya wrote that she had seen a large flock of Amur falcons in the rainforest while looking for hornbills. About the size of a pigeon, these small birds of prey are the same grey colour as city pigeons, but their darker heads and marks around the eyes give them the appearance of motorcyclists wearing helmets and goggles. Amur falcons make one of the longest migrations in the world, travelling an astonishing 8,000 miles within a few weeks, and along the way gathering in large flocks of several thousands in the Indian state of Nagaland. The spectacle of thousands of falcons flocking together is the bird equivalent of the wildebeest migration in the Serengeti. At this time of year, the falcons were on the move from their breeding sites in the Amur region of east Russia and Mongolia to their wintering grounds in the south of Africa. I wrote back saying that I had seen a flock of Amur falcons a couple of days ago in the deserts of Mongolia. Perhaps we had seen the same flock? This felt like the most romantic thing that could happen to a wildlife biologist living alone amidst the sand dunes of the Gobi Desert.

Another month went by and my fieldwork was over; then I spent two months working in a genetics laboratory at the National University in Ulaanbataar. I was ready to return to India and get married. During this time in the Gobi Desert, the species I had not given much thought to was the humble domestic goat that produced world-famous cashmere wool. On reaching Ulaanbaatar, however, I planned on buying Bhagya a cashmere scarf. Only when I entered a souvenir shop did I realize that it would cost me six months of my PhD stipend, so I settled on a pair of gloves which cost half that. Little did I know that this tryst with cashmere would dominate the next decade of my life.

The word cashmere comes from the word Kashmir, a province in

Northern India. This region has produced some of the softest and warmest hand-woven and hand-spun fabric in the world for the last 700 years. When the Western world experienced this fine fabric and its intricate embroidery, it came to be known as cashmere after the name of the region. But it is a misnomer, because the wool of this exquisite fabric is not produced in Kashmir. For several hundred years until the beginning of the nineteenth century, all the world's cashmere was produced in a single place: the Changthang plateau of western Tibet. This plateau has an average elevation of over 4,500 metres and spreads over 500,000 square kilometres, making it a little larger than Spain. The extreme dry and cold weather led to the development of a fine fleece in the local breed of goats. The Changpa herders would shear this undercoat and trade it with the people in Kashmir in exchange for grain, tea and other things that don't grow at high elevations. In Kashmir, the fine undercoat would then be cleaned, hand-spun and woven by artisans to produce the regal cashmere fabric.

Locally, in Kashmir, cashmere is called pashmina, after the Persian word *pashm* meaning soft gold. However, the prize for the finest fibre in the world goes to shahtoosh, made from the wool of the Tibetan antelope (*Pantholops hodgsonii*), locally called chiru, which occurs in the same regions of western Tibet where pashmina is produced. Desire for shahtoosh brought the Tibetan antelope to the verge of extinction in the twentieth century.

The name Tibetan antelope is another misnomer. The animal is not a member of the Antelopinae sub-family which includes all the world's antelopes. It belongs to the Caprinae sub-family of animals including goats and sheep, as well as the ibex, the blue sheep and the argali. But it looks nothing like a goat or a sheep, instead more closely resembling an antelope with a stocky body, supple legs and two long, straight horns that stick out like the two stems of the letter 'V'. They live in large herds sometimes reaching 600 strong and migrate large distances of 300 to 500 kilometres in the high pastures. The males and females graze in separate areas, coming together only during the breeding season in autumn. Females congregate in large numbers for calving and some of the most important calving grounds were discovered only in the early years of the twenty-first century.

A single chiru has only about 125 grams of shahtoosh on its body. The word shahtoosh is a combination of *shah* or king and *toosh* or down/fur; this is the king of soft textile fibres. The chiru population, estimated at around a million individuals at the beginning of the twentieth century, had plummeted to a mere 75,000 by the 1990s. At the peak of its extraction, an average of 20,000 chiru were hunted each year, indiscriminately shot with automatic weapons in the headlights of ATVs. Even the large herds of calving females were targeted.

Once upon a time, waves of chiru herds would graze parts of the Tibetan plateau in China, India and Nepal. Today, chiru is all but extinct from Nepal and India and only found in parts of Tibet, Xinjiang and the Qinghai province of China. Occasional herds move across the international border between China and India in western Tibet.

Almost the entire shahtoosh haul obtained from this carnage would make it to Kashmir through traders in Ladakh. The skilled artisans of Kashmir turned it into one of the finest pieces of fabric seen by humans, fetching up to $17,000 for a scarf or a shawl in Western markets in the 1990s. Because of the large sums of money involved, shahtoosh represented a conservation challenge of epic proportions, but since it went through a bottleneck in Kashmir, governments and NGOs were able to tackle it. In 2001 and 2002, conservation activists from the Wildlife Trust of India secured judicial and legislative decisions to ban all trade in chiru, its body parts and their derivatives in the state of Jammu and Kashmir, bringing the entire shahtoosh shawl industry to its knees. While this slowed down the hunting of this magnificent animal in parts of Tibet, it rendered over 10,000 highly skilled artisans in Kashmir unemployed.

Pashmina was the next best fibre that could be exported from the Changthang plateau to Kashmir to save the Kashmiri artisans. At least for the moment. The pashmina obtained from the goats of Changthang was already an established luxury product of the region.

Before automatic rifles and ATVs, it had been hard to procure large amounts of shahtoosh, and only about 200 kilograms of shahtoosh (produced by 1,600 chiru) was being brought annually to Kashmir by the second quarter of the twentieth century. Compared to this, more than a century earlier, in 1820, already an estimated 40,000 kilograms of pashmina was coming to Kashmir from western Tibet. As early as the sixteenth century, Kashmir was exporting cashmere shawls as far away as Iran. By the late 1700s, cashmere had made its mark on British and French fashion. Empress Josephine, the first wife of Napoleon and a fashion icon, can be seen draped in cashmere shawls in multiple paintings. She is believed to have owned over 300 cashmere shawls costing more than 20,000 gold francs.

That I could buy a pair of cashmere gloves for Bhagya made me feel like an attentive fiancé. I returned to India in early 2012 and we were married soon after. Bhagya did not get the opportunity to wear the gloves in the hot Indian summer that followed. We had both completely forgotten about them until Charu, my PhD advisor, shared a research manuscript with me. Led by an American professor and a Mongolian researcher together with Charu, this study highlighted the consequences of the globalization of cashmere for local biodiversity. While all the attempts of French, British and American entrepreneurs of the nineteenth and twentieth centuries to raise cashmere goats in Europe, the Americas, Australia and New Zealand ended in bitter failure, herders and traders were successful in raising these goats in the arid and cold climates of Mongolia and the high mountains of Central Asia. By the year 2000, about 40 per cent of world's cashmere was being produced in Mongolia and another 35 per cent in China, while the remaining 10 per cent was divided between the smaller central Asian countries, Iran and India.

The areas where cashmere is being produced map exactly on top of the high mountains of Asia, and the global range of the snow leopards. With 90 per cent of this cashmere exported to Europe and North America and with strong Western demand, the price of cashmere has increased and created an economic incentive for herders to develop the goat population. A kilogram of raw cashmere in India increased from 1,200 rupees in 2008 to over 4,000 rupees in

2022. Scientists traced two direct consequences of this shift. Firstly, that herders who would otherwise keep a diversity of livestock species such as sheep, goats, cows, horses, donkeys, yaks, Bactrian camels and cow-yak hybrids were now reducing all other species of livestock and herding goats alone. The grazing of a rangeland by multiple species of herbivores is better than grazing by a single species. A diverse set of herbivores spreads the grazing pressure across a range of plants. Intense grazing by a single species can be detrimental to pasture health as they tend to completely eat up palatable plants and leave only the ones that they cannot eat. Secondly, goats are better competitors of wild herbivores. More than eight species of threatened and endangered wild herbivores were suppressed by increasing populations of cashmere goats. Some of the most affected species were the chiru, the Tibetan gazelle, the Bactrian camel, the takhi (wild horse), wild yak, blue sheep and ibex.

My research had already shown that wild prey was key to healthy snow leopard populations. Now we discovered that cashmere goats suppressed the populations of wild herbivores. This was bad news for snow leopard conservation. And with the increased economic importance of goats, herders were less tolerant of snow leopards eating their prized cashmere goats and would retaliate swiftly. The research paper which Charu shared with me ended with a call to make the cashmere industry more sustainable and friendly for wildlife conservation in the region.

But the scale of response needed was mind-boggling. Increased demand for cashmere from Europe and North America was rapidly changing how local people herded their livestock across the high mountains of Asia, with a profound impact on the rangelands, its wild herbivores and the snow leopards that live there and nowhere else on the planet. Mongolia alone had about 60 million cashmere goats in 2012. A ballpark estimate put the total number of goats in high Asia at over 150 million. They were part of the fabric of life on this harsh roof of the world, and cashmere was – and remains today – the primary livelihood for the tens of thousands of herders who live here.

*

One night in the desert, I had finished my chores and was ready to get into my sleeping bag, when I heard a car engine. Ears pricked, I stood waiting to hear which way the vehicle went. It kept getting louder and then stopped outside my *ger*. I had never had visitors this late in the night. As expected, someone pushed the door. The herders of South Gobi never latch the door of their *ger* and nobody ever knocks. You just walk in and trust that the residents heard you coming.

It was Midjii and his father-in-law, Baatar. Baatar was tall and he looked even bigger than usual in his heavy maroon-coloured *deel* – a traditional wraparound dress made of wool – and a warm, fur-lined cap. Their faces were red from the cold winds. '*Saine bain oh*,' ('how are you?' or more like 'hello') Midjii muttered under his breath. I pointed towards two wooden stools and I made a move to heat water to prepare tea. But they did not sit; there was a sense of urgency in their movements.

'*Irbis*,' Midjii said, gesturing for me to follow, and left without waiting for me to dress for the cold night. *Irbis* meant snow leopard in Mongolian. They were leading me to a snow leopard, I figured. Glad that I hadn't dressed for bed, I grabbed a down-feather jacket and a woollen hat and stepped out after Baatar. Midjii was ready at the steering wheel of his grey Russian van.

The night was very quiet, and it was hard to see anything beyond the circle of the headlights. I felt very alert. Where were we going? Was it a dead snow leopard? When we reached Baatar's camp, Midjii went towards the *ger* but Baatar led me towards his corral, where he had over 300 cashmere goats. We went around the pen and Baatar crouched low as he came to the crest of a gentle rise. Baatar sat down and, pointing into the dark night, repeated, '*Irbis*.'

Lying down beside him on a mud bank, I peered into the dark. I could not see anything. I pulled a torch from my coat pocket and signalled to Baatar to ask if it was okay to use it. He nodded. I aimed in the direction Baatar had pointed and switched it on. Two orange eyes were staring right back at me. Although it was hard to judge the distance, it could not have been more than 40 metres away. I pulled out my binoculars and lined them up with the torch to see more details. That was when the goriness of the scene became visible.

A snow leopard was eating the carcass of a snow-white cashmere goat, and there were at least three or four other dead goats. The snow leopard kept looking back at us while tearing at the red flesh.

Baatar was facing me, waiting to see my expression. I offered him the binoculars but he refused and then he stood up and walked straight towards the snow leopard with his hands in the air, shouting as if he were herding goats. I got up in a hurry and tried to keep up. A few seconds later, I mimicked him in shooing away the snow leopard. The leopard dragged away a goat and disappeared into the dark hillside. It was only the next morning that we could assess the full extent of the damage.

The herders were showing me what they had to live through. What it felt like to be at the receiving end of the snow leopard's aggression. What it is like to live alongside this big cat.

But cashmere is not the only product reshaping the snow leopard's habitat. At the turn of the century the Mongolian economy was undergoing a tectonic shift. It was among the fastest growing economies in Asia, due to its government's focus on mining for coal, gold and other minerals. My colleagues in Mongolia were shocked to learn that around 10,000 square kilometres of land around our base camp in the Gobi Desert – the entire territory for Aztai, Khasha and her cubs, and around twenty other snow leopards – had been given away to mining companies for prospecting. This place was teeming with numerous ibex and argali. Around seventy herders with thousands of cashmere goats have called it home for generations. The Mongolian country director for the Snow Leopard Trust, Bayara (the same person who had started the Snow Leopard Enterprises for women) immediately began to work on a response.

Although exploratory mining licences had been given, mining had not yet started in the region. Bayara used this time to talk to the local herders. We were sure that the mining companies would lure the locals with jobs and offer financial compensation for their grazing lands. But the herders wanted to continue their way of life. Mining was not only a threat to wildlife but also to the local cashmere-herding economy. The herders were earning well from selling cashmere and were their own bosses. The free-spirited people riding their horses and motorcycles over the open Mongolian steppe were not interested in the jobs that the mining companies were likely to offer. Bayara found that under the Mongolian law, people with traditional land rights could create a nature reserve on their land if they could prove the biodiversity value of the land and if they could fund the management of the nature reserve. Bayara worked with the herders and made a plan: her team would raise the necessary funds for managing the park while together the herders and young researchers would collect information to prove the biodiversity value of the region. Satellite telemetry data was already available to prove that this was an important habitat for snow leopards. There was already some documentation for plants and birds. We needed information on the herbivores: the ibex and the argali.

I returned to the south of Gobi the following autumn. This time, the plan was to work with a young Mongolian student to help him set up a study on the diet of the ibex and train him to use the Double Observer Survey to estimate the population of the herbivores in the larger region. I met Lkhagvasumberel Tumursukh, or Sumbee as everyone called him, at the Snow Leopard Trust's base camp in the Tost mountains. Sumbee had set up his own *ger* next to the camp and he was also appointed as a manager and researcher of the long-term research station.

Sumbee was built like a wrestler and liked to wrestle his friends during free time. He was tall, broad, with strong shoulders and a round baby face with high cheekbones. He would tower over you, but also look at you in an affectionate way. He was very competitive. He had to be the first one to reach the top of a mountain, the fastest on the motorbike, carry the heaviest load, and he was unabashed about it. He was passionate about conserving the wildlife of the Gobi. I had seen pictures of him carrying heavy barrels of water strapped to his muscular shoulders uphill on mountains to replenish the waterholes during a drought earlier in the summer. This back-breaking work was not part of his responsibilities. It seemed futile to fill waterholes in the remote desert one barrel at a time but Sumbee used all his free time doing just that until animals started showing up at these places. This spirit would come in handy for his research on the ibex and argali.

I spent three weeks with Sumbee and every morning we would locate a herd of ibex or argali and I would teach him everything that I had learned from my PhD advisors, Sushil, Thinley and the people of Tashigang and Kibber. Sumbee practised the Double Observer Survey and loved watching the ibex. I was happy re-living my days studying the foraging of the blue sheep in Tashigang.

A little over a year later, the hard work was paying off. Sumbee completed his data collection and visited me in India to write his master's thesis. Before he returned to Mongolia, Bhagya and I drove him to Bandipur National Park. We were excited to see elephants, bisons and dholes (wild dogs). In the evening when we were driving back to our cottage a bull elephant charged our car and stopped only a few feet away from us. I looked at Sumbee and he smiled back, unfazed. He was warm in a childlike way and afraid to show his emotions or fear, be it of academics or elephants. What led him to drop his guard was a peacock displaying his colourful train. Sumbee gleefully hopped out of the car and took hundreds of pictures. He kept looking at the peacock and then back at us like an excited kid. At a loss for words, he turned back and took more pictures of the beautiful bird.

By the time Sumbee returned to Mongolia, Bayara, the country director of the Snow Leopard Trust, had helped the herders campaign for the land and local officials had declared that an area of about 7,500 square kilometres would become a nature reserve. It was to be called the Tost-Tosonbumba Nature Reserve, named after the two mountain ranges in the region. The herders had drafted the laws of the nature reserve where they would continue to graze the lands but also work as rangers to prevent illegal hunting. The Snow Leopard Trust would provide a salary to these herder-rangers. When I asked Bayara if this meant that the future of snow leopards was secure she explained that the reserve was approved only at the *soum* (county) administration. A lot more work would be needed before it became a nature reserve at the national level. 'Only with national protections can we claim that this place is secured,' she told me.

Once Tost-Tosonbumba was declared a county-level reserve, our aim was to find ways to create harmony between snow leopards and herders. Although cashmere goats competed with wild herbivores for forage, they did not pose an existential threat to snow leopards. Mining on the other hand could be a death blow. For cashmere herders to support conservation and oppose mining, livestock predation by snow leopards had to stop or be significantly reduced. Together with Sumbee and colleagues from Mongolia, Bayara launched a study to assess the feasibility of making the corrals in the landscape predator-proof. We visited many herders to understand the level of livestock predation they faced and to estimate how much work it would take to secure their corrals. Herders in the Gobi have 400 to 600 goats; sometimes reaching 1,000. They used motorbikes and horses to herd their livestock and guard dogs whenever possible. They moved between the open steppe desert in the summer and the hilly Tost-Tosonbumba mountains in the winter. Typically one herding family had two or three camps in the mountains and one or two camps in the steppe. The mountains shield them from the cold Siberian winds that descend from the north during the winter, but the camps there were more vulnerable to snow leopard attacks. At the winter camp, the herders barely had a corral. Usually, the goats were gathered in the shelter of a simple wall reaching a height of 2 to 3 feet shaped like the letter C with a narrow opening on one side. Sometimes, a small part of it was covered with a roof. It was very easy for snow leopards to hop over the wall and pick out a goat. The herders had dogs, but not all dogs were equally good at their job. The open structure meant that snow leopards did not get trapped inside these corrals, but when a snow leopard visited a corral repeatedly, the herder would deploy foot snares or sit at the corral with a gun.

Killing a snow leopard, even in defence of the livestock, is illegal in Mongolia (as it is in all twelve of the snow leopard's range countries). But Mongolia's political history had negatively affected the relationship between herders and snow leopards. In 1921, Mongolia became only the second country in the world to embrace communism. For the next seven decades, it remained in the shadow of Russia on the one hand and China on the other – the only two countries with which Mongolia shares its borders. During the time of communism, carnivores like snow leopards and wolves were declared enemies of the state, and herders were organized into teams to hunt them. Awards were given for the most successful teams and the most successful hunters. Hunting snow leopards brought social prestige and government appreciation. I once met an old herder who had killed four snow leopards when he was young. Back then, he said, 'I was a hero, a local legend.' Today, in his old age, he tries to take pictures of wildlife with an old camera gifted to him by some tourists. During communist times, herders acquired tremendous natural history skills learning how to keep snow leopards and wolves at bay.

In 1990, Mongolia transitioned from communism to a multi-party democracy and started privatizing overnight. Livestock, which was communal property, was re-distributed and became private property. Everyone was now required to protect their livestock, and the state was not giving any prizes for hunting carnivores. Herders continued to hunt snow leopards and wolves. The new government passed a series of laws related to the protection of the environment, the creation of nature reserves and the management of wild animals in the years 1994 and 1995, and while these laws made it illegal to hunt snow leopards, they were not implemented for a long time. For a newly formed country that had recently weaned itself from the support from the USSR, animal protection was not the priority. It was another decade before aid from international organizations helped set up a series of nature reserves and the environmental protection laws were implemented. Herders accepted this change grudgingly. About half of Mongolian people herd livestock and therefore accepting, even protecting, the predator of their livestock met with resistance. The wolf still remains unprotected outside the nature reserves in most provinces, encouraging hunters in those regions and thus continuing to threaten the snow leopards.

One winter, when a herder in the western part of the Gobi Desert was repeatedly troubled by a carnivore killing his goats, he decided to set some foot snares around his corral. When a commotion woke him up in the middle of the night, he stepped out with his gun. He pointed a torch towards the snares and saw an eye shine. He aimed and shot and ended the light in those eyes forever. When he went up to what he thought would be a dead wolf, he was disturbed to find a snow leopard with a collar around his neck. 'That's how Bayartai, the big male, met his end,' Örjan told me. The herder said he thought it was a wolf that was killing his goats. He had not confirmed the identity of the predator in the dark before he fired the shot. He did not want to get too close because he feared being attacked by other wolves.

These difficult interactions with the wild carnivores that preyed on their livestock meant that it was up to us to help herders protect their livestock when they supported the creation of a nature reserve. With Sumbee and other colleagues from Mongolia and Sweden we mapped out almost all the corrals across the Tost-Tosonbumba mountains. We identified certain corrals that were more vulnerable because they were close to rocky areas and cliffs from where a snow leopard could stalk the goats. We interviewed the herders to understand the baseline levels of livestock predation by snow leopards and wolves in these corrals. During the interviews, we studied the attitudes of herders towards snow leopards and wolves. The plan was to help herders make predator-proof corrals and then estimate whether a reduced level of livestock predation improved herders' attitudes towards these carnivores.

But we did not have the means to make all the corrals predator-proof. There were over a hundred corrals. In the first phase, we had the time and resources to build ten. As scientists, we wanted to conduct a randomized and controlled trial, so that no factors other than the ones being tested would influence the outcome of the experiment. The first step in our study would be to identify the treatment group, the herders whose corrals would be made predator-proof, and the control group, whose corrals would remain the same. We were unsure whether herders would accept such a study design, as they might want the most vulnerable corrals to be made safe first. That would be fair but it would not help us with the research. Bayara, the country director, came to our rescue. Bayara commanded immense respect from herders across this region due to her decades of conservation efforts. She led a meeting of the herders where they created a group of people whose corrals were vulnerable and then she explained the study design and the herders agreed that within the group of vulnerable corrals, the treatment and control group could be chosen at random.

Ten corrals were identified for the treatment group, and ten were identified for the control group, and the project team immediately got to work. Using our combined expertise on snow leopard and wolf behaviour, access to building materials in Mongolia and the local conditions, we came up with a design. We planned to create rectangular corrals, two metres high, with chain-link fences, metal poles and an electric wire at the top. The body of the fence would not be electric, so if a child or the goats touched the fence, they would not feel anything, but if an animal tried to climb the two-metre fence and came in contact with the top wire while being in touch with the rest of the fence then the climber would receive a high-voltage pulse. The electric wire would be powered by a car battery connected to two solar panels, which would be positioned at the *ger* of the herder.

We got to work. Our camp looked like a construction site. Heavy equipment was brought in to cut metal poles and the chain-link fence. We transported the material in a trailer behind Midjii's Russian van to the herders' camp, which was more than 50 kilometres across loose gravel and sand. After all the measuring and planning, the first task was to hammer the metal poles into the ground, and Sumbee took charge. He used his wrestling strength to get it done quickly. The herder family worked alongside us. The quiet little corner of the desert became a bustle of activity. Every member was busy hammering, cutting, tying, mixing cement or cooking for the team.

After we completed the first corral and we stood inside the two-metre-tall fence with its protective electric wire, I remember someone joking, 'Let alone snow leopards and wolves, this fence would stop a velociraptor from taking the goats.' I was proud to be part of the team that was directly contributing to the well-being of the herders in this faraway corner of the world.

As the sun set, a golden light spread horizontally over the desert with hues of the sand and rocks. Everything around us, the rocks, the herders' *ger*, the goats with a layer of dust on them, the guard dogs, the stone-grey Russian van, blended into the light. The only thing that stood out was the freshly made steel corral. It was an alien structure in the landscape. I wondered how I looked too. Was I blending in, or was I standing out?

A year later, when the results of our randomized controlled trials came out, we had mixed results. Not a single predator-proof corral lost a goat to snow leopards. In fact, the herders said that the electric wire was unnecessary. The wolves and snow leopards don't come close to the structure, let alone climb over it. Sumbee was especially proud. We were relieved that this effort would help protect the livelihoods of herders while also protecting other snow leopards from meeting Bayartai's fate.

But when I looked at the interview data and examined the attitudes of the herders, something was off. The results showed that making the corrals had not improved herders' attitudes towards the snow leopards or the wolves. The ultimate goal of most conservation projects trying to reduce the impacts of wildlife on local communities is to promote harmonious coexistence. Unless local people have a positive attitude towards snow leopards, they would not tolerate their presence. Effective predator-proof corrals would ensure a reduction in incidents like the one that killed Bayartai, but for conservation to succeed in the long-term and across large areas, we had to find a way to change people's attitudes.

Over the coming years I continued to work with Sumbee on other projects. Bayara had made steady progress and her team celebrated the reserve being approved at the provincial level. We had unprecedented data on the snow leopard from the Tost-Tosonbumba mountains of the Gobi. Sumbee had estimated the population of ibex and argali throughout the mountain range and the surrounding mountains. Another team, together with Sumbee, had used camera traps and found that thirteen snow leopards lived here and over a period of four years, twenty-one young had been born. Despite the data, persuading the Mongolian government to protect this region from mining at the national level was harder than we first thought.

One morning I was working in our office in Bangalore when I received a phone call from Mongolia. I had not been back in many months and I was excited to hear from my friends there. I was not prepared for what followed.

'Sumbee has been missing for the past two days and nobody has heard from him during this time. His phone is switched off.'

I did not know what to make of it. Maybe he wanted a break and had taken off somewhere, but he was not the kind of guy who would leave without informing anyone. I thought maybe he eloped with his girlfriend, but none of his friends or colleagues knew of anyone he was seeing. The caller said they would keep me posted and hung up.

I could not work for the rest of the day as I anxiously waited to hear from Mongolia. There was no news for a couple of days. The police had been informed and they were looking at CCTV footage around the places where Sumbee was last seen. His disappearance was local news. Some were talking about the possible involvement of the mining companies. Sumbee had had confrontations with them over the protection of the Tost-Tosonbumba mountains.

I was trying to remain positive and not think of anything extreme. I also felt extremely helpless. I wasn't able to leave for Mongolia immediately, nor was I sure how I could help, since I did not speak Mongolian or Russian and our team in Mongolia was doing everything they could.

A couple more days passed and we were getting desperate. I heard that the police had found a recording of an SOS message that Sumbee had sent to his sister but it was hard to know what was true and what was rumour.

When I picked up the next phone call from Mongolia, the caller was in tears. 'They found his body in Lake Khövsgöl near his parents' house. There are no other details.'

I was in shock. I called up a few other people in Mongolia who were similarly distraught. I left the office and went home. Bhagya was as shocked but she was stronger.

Sumbee's death became an international news story. There was no clarity over what had happened but the possibility of the involvement of mining companies was adding to pressure on the national government. International conservation organizations sent letters to the prime minister of Mongolia.

Sumbee was last seen on 5 November 2015 and his body was found on 11 November. He was just 27 years old when he died. We still don't know what really happened. What we know is that Sumbee loved the ibex, the argali, his study subjects, and the snow leopards. To this day, Bhagya and I think of Sumbee when we see a peacock displaying his train.

Bayara continued her efforts to push the local authorities to find answers about Sumbee's death. She also redoubled her efforts to have the Tost-Tosonbumba mountains, the place Sumbee had dedicated his life to, declared as a state nature reserve. Five months later, in April 2016, news came that the Mongolian Parliament had approved the Tost-Tosonbumba Nature Reserve. Bayara won the prestigious Goldman Environmental Prize in 2019 for her efforts to protect the Tost mountains. The campaign led to the cancellation of thirty-seven mining licences.

The dedication with which Sumbee would carry water on his back to refill the waterhole during droughts in the Gobi Desert was a testament to his passion for conservation. Nine years after his passing, Sumbee's Mongolian colleagues published two research papers, the first one showing that snow leopards dig into waterholes during extreme droughts and this water becomes accessible to a whole suite of species in the desert. Snow leopards are like ecosystem engineers, making the desert a more hospitable place for multiple species. The second article described a female snow leopard with a one-year-old cub visiting a waterhole thirty-seven times over a period of sixty days during the summer. What is clear from these two reports is that availability of water affects where snow leopards can live in desert ecosystems like the Gobi. Those of us working in the Himalaya do not think about it much because the mountains are full of glaciers, snow and streams. Sumbee had known it before any of us and was already working on the problem in his own way by using his strong body to replenish as many waterholes as possible.

Ecology taught me that the cashmere-producing goats can degrade the pastures and even suppress the populations of blue sheep, ibex, argali and other important wild herbivores, and my time in Mongolia taught me that the herders who raised these cashmere goats were the most important custodians of the mountain lands. Herders and snow leopards had shown great resilience in living alongside each other. Without the herders, the mountains of Tost and Tosombumba would have been hollow mining pits in the earth. Cashmere production keeps the land in its original form of a rangeland where wild herbivores and carnivores can live alongside the herders. Cashmere production needs to be managed, its ecological footprint needs to be limited and, in return, it ensures vast swaths of land are kept as rangelands where other species of wildlife can also coexist. Without the herders, the places would change completely. Whether it is mining, a solar power project or other infrastructure development, these land uses transform the place beyond recognition. There is no room for the ibex, blue sheep, argali or snow leopards in these new schemes.

The fates of the snow leopard and its prey species are deeply entangled with the fates of the herders and their cashmere goats.

# 7

# Crooked Corrals

Tomden lived in a beautiful two-storey mud house in the Spiti Valley. Just beyond the garden in front of the house, the Lingti River gushed in a series of rapids. At the back of the house were fields where he grew barley and green peas. A series of small gutter-like channels along the contours of the mountainside brought fresh water from the Lingti to irrigate his fields. A small stream beyond the fields joined the Lingti and a small wooden bridge over the stream connected his house to the walking path beyond. A gate made of driftwood at the end of the bridge kept animals from grazing his fields. On both sides, snow-capped mountains rose to towering heights of 5,000 metres, over a thousand metres above his house. Tomden lived in this Eden with his wife and two kids, their fifty goats and two cows. His only neighbour spent most of his time in the administrative town of Kaza. Tomden's wife was a schoolteacher in Kaza and spent most working days there. His children went to a boarding school but during their winter and summer holidays they herded the goats with their father.

One spring evening, Tomden returned home from the mountains with his goats. Their bleating became louder as they got closer to the house. He opened the wooden gate and let them in one by one, counting as they ran inside. The two cows ambled along at the back. It was dusk by the time they drank water from the trough and he put them inside their pen on the ground floor of his house. He had three rooms: one for the big goats, one for the small ones and one for the cows. They each had a small window for ventilation to allow the methane released by their farts to escape. The escaping gases would pull a draught of fresh air through the gaps in the door. The pen was warm on cold nights.

Tomden had a smoke on his porch before making some noodle soup *thentuk* and going to bed while listening to the radio.

He did not notice at first but the next morning was unusually quiet. He made himself some tea, added some roasted barley flour *tsampa* to it and had it for breakfast. When he went downstairs to open the door to the three pens, he realized that the goats were not restless to go out. He became alert. He brought a torch and shone it through the slats in the door and saw two eyes shining back from the corner of the pen. The eyes of the dead goats did not reflect back his torchlight. Tomden's blissful life was shattered.

He released the young goats and the cows from the two other pens next door and did what he had to do to avenge his loss.

Thinley and I parked in the neighbouring village and walked an hour to Tomden's house. The path descended down a steep mountainside before following the stream to the wooden bridge leading up to the driftwood gate. Pink wild roses were in bloom along the stream. Tomden warmly welcomed us at the gate. He had just returned home from the mountains with his goats and cows. Tomden said he had been expecting us for a few days. Time moved at a different pace here. He was in no rush and that calmed us down. Thinley introduced us as the people he had spoken about during his previous meeting with Tomden in Kaza. Tomden shook our hands and invited us inside. We wanted to soak in a little more of the beautiful surroundings but we could not refuse his invitation of tea. As we settled down inside the house, he narrated the story of the day the snow leopard entered the goat pen and killed thirty of his flock. The next morning, the snow leopard was so full that he could not climb to the small window through which he'd entered, which is higher up the wall from inside the pen than it is from the outside. After tea, we went to see the pens and the windows.

'Are these goats the survivors from that night?' I asked naively.

'No, he killed all the big goats. I was left with only the kids and the juveniles but people from the neighbouring village traded a live goat for a dead one each. Everyone needs to butcher a goat for meat every few months so they accepted my dead animals. Most of them were untouched by the snow leopard anyway. Many of them died in the stampede when the snow leopard got in. The few large ones that survived were killed by the snow leopard in a hunting frenzy. But he ate only one.

'I don't know how I slept though that stampede happening right below where I lay that night. I dispatched the snow leopard that morning and threw him in the river,' he said without guilt, shame or anger. This was part of life. I did not ask him how he did it. It felt too intrusive, even though I was very curious. Tomden had maintained a dignity throughout the story which would be broken if I asked this question. Tomden knew –Thinley had clarified it to him – that we were not meeting with the idea of proving guilt. We wanted to understand the situation and ensure that such an incident did not repeat itself.

After the initial awkwardness had passed, Tomden entertained us with stories of snow leopard sightings. He told us that we should come back in the winter and we would surely be able to see a snow leopard from his house. He pointed to a rocky promontory a little higher up on the mountainside where one liked to sit on sunny mornings and soak in the warmth.

He told us the story of how one day a blue sheep kid joined his goat herd. He tried to shoo her away but she would not go. The blue sheep came back home with them and started living with his goats. She grew up very fast and soon was one of the largest in the herd. The blue sheep lived with them for more than two years and then one day, when his goats were grazing close to a herd of blue sheep, she seemed to have an epiphany and walked over to the blue sheep herd, joining them and never looking back once.

The sun was behind the mountains and a cold evening wind had picked up. Thinley added cakes of dried cow dung to the metal stove. We asked Tomden what he thought about covering the small windows of his corrals with a metal grille so that the snow leopard could not go inside. We would split the expenses of the metal grille and the fixing between us, so that Tomden would not have to pay the full amount. Tomden nodded in agreement but did not seem too excited. He said to Thinley that finding a metal welder and getting it fixed properly was a time-consuming task and he could not spare any time away from his animals. They need to be taken to the pastures every day. 'We can do that for you,' Thinley suggested. Tomden was happy with the arrangement, and thought that his problems with the snow leopard would be solved if we were able to do this. It was a simple solution to a simple problem, although it took Thinley and me a week to find the metal-works guy in Kaza. He was overburdened with pre-orders. We pleaded with him and paid him extra to expedite our request. We did not want another incident of a snow leopard entering the corral.

It was getting dark by the time we ended our conversation and Tomden suggested that we stay the night at his place. The thought was tempting. It was a beautiful place and Tomden had incredible stories to tell, but our other colleagues did not know our plans and they would be worried if we did not make it back to our base camp by night. We promised that we would stay the night when we come back to fix the grille on the corral windows.

The entangled lives of the snow leopards, their prey, the herders and their livestock leads to give and take every once in a while. The herders understand this better than anyone else. Most herders are tolerant of a snow leopard stealing a goat from the pastures occasion-ally. Often, the snow leopard will pick off an injured or otherwise unwell individual. Some herding cultures even welcome the death because they believe that such predation keeps the herd healthy and prevents the spread of diseases.

However, snow leopards killing livestock inside the night-time corrals and the herders' retaliatory persecution were a real problem, first identified by biologists and conservationists in the 1980s and the 1990s. It was like an earthquake that shook the ground beneath your feet and brought down the house. Scientists had conducted programmes to make herder corrals 'predator-proof'; built so well that snow leopards could not enter them. The electrified corrals we built in Mongolia were one such effort.

Keeping the snow leopard out of a corral is not an easy task. We know from research on captive snow leopards that they have fairly advanced cognitive abilities. When presented with puzzle boxes, snow leopards are able to find multiple solutions to a problem. There is also a fair amount of difference across individual animals. Not all snow leopards follow the same approach to finding solutions, and not every snow leopard is equally motivated. The main predictor of a snow leopard's ability to solve a puzzle is their persistence. If an individual animal is motivated and persists with the problem in front of them, chances are they will solve it. When a snow leopard has set their mind on getting into a corral (with easy livestock prey inside) it's likely they will find a way. The corrals not only have to stand against the strength of the snow leopard but also survive the intelligence of the animal. No wonder the snow leopard had killed so many of Tomden's animals without him being alerted to it. I believe the snow leopard knew that he should not disturb the herder who was sleeping not too far from the goats.

Even when we find a way to secure a corral and keep the snow leopard away, scaling is expensive and the risk of livestock predation by snow leopards is not uniform. Security varies due to the location of the corral in the surrounding mountains, the population of blue sheep and other wild prey in the region, and the behaviour of the snow leopards in the region. We have also found that herding practices in different regions of the snow leopard's distribution range vary. At fifty goats, Tomden had one of the largest flocks in Spiti. Here, people owned few but diverse species of livestock. Average families in Spiti would have one or two yaks that they used for ploughing, a horse or two for ceremonial use, two to four donkeys to carry their agricultural produce from the fields to the road, a few cows and cow-yak hybrids for milk and dairy, and some goats and sheep for meat. Spitians live in settled villages and do not keep dogs. In fact, many villages have laws against dog ownership because dogs often became feral and kill the livestock they are supposed to protect.

But this region is also the home to groups of herders from other parts of the Himalaya. The Gaddi, the Bakarwal, the Gujjar and others make annual treks of hundreds of kilometres from the base of the Himalaya, where they spend the winter in relative warmth with plenty of grass, to the high-elevation mountain steppes in summer. They travel over snow-covered passes above 5,000 metres with hundreds of goats and sheep, sometimes even thousands, and different herders band together to tackle high passes with deep snow. Some of these herders have traditional grazing rights to these faraway lands; others pay a fee to the local villages to access their pasture lands. They are constantly on the move, and making a predator-proof corral at every camping ground is not possible.

A report in 2016 based on information contributed by many snow leopard biologists from around the world concluded that an average of one snow leopard had been killed every day during the previous ten years. In most cases, the death was in retaliation for livestock killing by the snow leopards, like the one at Tomden's house. Once the snow leopard had been killed, their body parts and skin make it to the illegal markets, creating a supply-side-driven trade in the species. This could create a demand for the body parts of the snow leopard leading to more targeted poaching to fulfil the demands of the illegal markets. Conservationists have witnessed this happening to tigers, where a growing demand for tiger body parts from East Asia has led to local extinctions in two well-known tiger reserves in India and a widespread decline in tiger numbers across the tiger's range in South and South East Asia. Strict law enforcement has led to some recovery in tiger numbers in parts of India and Nepal, but the decline continues unabated everywhere else. We feared a repeat with snow leopards, and considering that snow leopards are found in much lower densities, their numbers could be threatened more easily. These concerns feed an urgent determination among conservationists to prevent snow leopards from getting killed in livestock corrals.

When the pandemic began in early 2020, I retreated to my ancestral village in central India. The world was coming to a grinding halt.

'Work from home' became the new mantra. But for herders like Tomden in Spiti, Baatar in the Gobi, or Tashi in Ladakh, this global event did not register. They were too isolated to be isolating further. Livestock herding could not be done from home, and livestock predation by snow leopards did not relent for a pandemic.

Sumdoo is a small settlement of seventy Tibetan refugee herders in Ladakh. These are herders from western Tibet who migrated to India around the time of the 1962 war between India and China when the border between the two countries closed down. *Sumdoo* means confluence, typically of two rivers. This beautiful settlement sits between two mountain streams at the isthmus where they merge together. The square-shaped mud houses have their backs to the cliffs overlooking the confluence. On the east, hills roll away in the distance, merging with the horizon. The vast openness of the landscape inspires a hope that one can see all the way to Lhasa, the capital of Tibet and the original home of Tibet's exiled spiritual and political leader, the Dalai Lama.

All the families in Sumdoo herd cashmere goats and produce some of the finest pashmina in the region. They live in the hope that one day they will return to their homeland in Tibet and continue their livestock grazing traditions. Even after sixty years of exile they have not accepted Indian citizenship. Every family owns large herds of cashmere goats. Small groups of herders join together in some pastures, and herders operate as individual families in others. In January 2020, the goats of most families were camping close to the village when a pack of wolves attacked one of their herds. The wolves killed a few animals, scattered the rest and drove them all over the plateau. It took the herders days to find the animals and bring them back. This was repeated a few weeks later, and then a snow leopard found a way into a corral and killed many goats. When the herders asked for government support in the form of compensation for their dead animals, the world was busy fighting Covid-19 and barely paid any attention. Within six months, their corrals were attacked a dozen times, and over one hundred animals were killed, amounting to over $10,000 in damage. When wolves and snow leopards started entering the corrals in Sumdoo in 2020, we were worried that the herders of Sumdoo would resort to desperate measures. During a brief respite from the lockdowns, a colleague drove to Sumdoo and shared the design of the corrals that we had built in Mongolia. This tried-and-tested design was sure to stop the wolves and the snow leopards but it did not impress the herders of Sumdoo. Their corrals were in a rockier terrain and they believed that snow leopards could jump in by climbing over nearby rocks.

Despite years of experience of building predator-proof corrals our team was hamstrung by the travel restrictions and Covid protocols. We could not risk sending someone to assess the situation and design a solution. And the situation in Sumdoo was getting more desperate. The longer we waited to make a decision, the greater the risk of the situation blowing out of proportion. If a herder ended up retaliating against a snow leopard, it might further alienate the people of Sumdoo from the local government. Meanwhile, the death toll from Covid in India kept climbing.

Just when we thought we were out of ideas, the herders of Sumdoo shared a hand-drawn sketch of a design for their corrals on WhatsApp that they believed would not only keep snow leopards and wolves out, but would also be suitable for the dry and cold weather of the high Himalaya. Rectangular in shape, it had 4-foot walls made of stones and mud. The two walls on the shorter side of the rectangle were shaped like the letter A, with a door on one side. The roof was like a tent sloping on either side and made of chain-link fencing, with a central pole to support it. The chain-link roof let in sunlight and breeze. The herders said that it should be aligned north to south to let in the maximum amount of sunlight to keep the corral dry and free of disease. The design was very simple, and other than a few chain-links the building material would be locally available, as were the necessary skills. The only thing that the herders needed was financial support, which we had already raised.

My big fear when working with herders to build predator-proof corrals is that we build a corral we believe to be predator-proof but then a snow leopard enters and kills livestock, or worse, gets killed in the process. A poorly built corral is like a lure for a snow leopard.

It risks snow leopard lives and herders' livelihoods and a conservation project should never make this mistake.

The herders had proposed something simple and local, and I was unsure it would work. Moreover, donors put their faith in high-technology solutions and sophisticated ideas proposed by experts. I would have to answer to our donors if something went wrong.

Despite these worries, we worked with the herders to implement their design. We made seven corrals for the herders who were camped in the most vulnerable locations. Two years later, when I could finally visit Sumdoo, not a single animal had been killed in any of the corrals designed by the herders. The herders proudly showed off the corrals that they had designed which had withstood the elements and saved their goats. An elderly woman said, 'When the goats leave for the pastures, I keep my valuables in the corral because it is more secure than my house.' She spoke warmly about the snow leopard, mentioning how beautiful and graceful it looks when it walks over the cliffs and rocks. I could sense a change in the way herders spoke about the snow leopard and the wolf.

'Nobody wants to harm another living being if they can help it. We want to live peacefully alongside all the creatures of the world,' another herder added.

Because of the pandemic, we had not been able to do a robust assessment of herder attitudes before and after making the corrals but I had no doubt that there had been a change in Sumdoo. I could not put my finger on what caused this change, when we had experienced no similar change in the herders of Mongolia.

After Sumdoo, we visited the neighbouring village Rupshu. The herders in this village claim – and many of the buyers support their claim – that they produce some of the finest pashmina in the world. I was accompanied by my colleague Ajay Bijoor. Ajay had joined our team in 2011 when I was still a PhD student. He had quit his corporate career with Tata Consultancy Services to join the world of wildlife conservation. He was very witty and always sported a smile under his bushy beard. We quickly became good friends and complemented each other in our work. I was interested in conservation science, Ajay in conservation management. We shared Marathi as our mother tongue. For the next decade or more that we worked together, we would unknowingly switch to speaking in Marathi when we were deep in thought.

We reached Rupshu in the approaching dusk. An endless carpet of green grass stretched ahead of us, only a few hills breaking the monotony. We were almost 5 kilometres above sea level. Somewhere in the distance, the border between India and China divides this plateau. Rupshu herders moved with their goats across the Himalaya between the turquoise waters of Tso Kar and the only highway that connects the town of Leh with the plains of India.

A large black Tibetan mastiff, locally called a *chanki* dog, stood guard over the pashmina goats while Ajay and I were deep in conversation with Acho Sonam. 'When it rains, the goats don't like to sit around in the open. They scatter in the hills to seek shelter under the boulders,' Sonam told me, 'and that's when the wolves pick them up.' My gaze tracked his to the surrounding hills, and I found myself imagining the piercing stares of hiding wolves, waiting for the rain.

Sonam had seen the designs of the corrals that we had built in Mongolia, Spiti and the neighbouring village of Sumdoo. But he was asking for something different. 'We need enclosures that can keep the goats *in*. We don't need to worry about keeping the wolves out; our dogs don't let them get close.' Sonam wanted a simpler but larger corral, through which the wind could blow more easily and keep the ground dry. Larger corrals are difficult to predator-proof, whereas the small and closed predator-proof corrals used on winter pastures are hard to keep dry. The moist floor of the corral is a breeding ground for disease.

As Sonam turned over the details of the corral and how it might keep the goats disease- and carnivore-free, it became obvious that he didn't like the corral design we had proposed. He wanted to build something different.

Again my challenge was in how to explain Sonam's wishes to my donors so far away, who would judge my work from the photographs that I would send. Conservationists and donors prefer NGO solutions to those developed with local people, and I knew that donors were likelier to fund a corral design conceptualized at a tech outfit in a glass office in Bangalore or London. We rarely think that herders might have simpler answers to preventing livestock predation by carnivores – but I had just witnessed it in Sumdoo.

'If you can help me build my corral to this design, I'll ensure that no snow leopards or wolves hunt any livestock and that none of us herders will touch any of these carnivores,' Sonam insisted. This was music to my ears; herders taking responsibility for snow leopard conservation. As a conservationist, what more could I want?

An evening calm was setting in when Ajay spoke softly in Marathi just for me to hear. 'The herders don't want a predator-proof corral; they want agency.' Agency to decide on the corral design, agency to manage their entangled lives with their animals and snow leopards. These herders have lived alongside snow leopards and wolves for thousands of years. All that time, they had the agency to manage their interactions with these carnivores. But snow leopards and wolves have become protected species since the Wild Life (Protection) Act of 1972. It took another couple of decades for the law to be enforced in Ladakh, but since then, herders have been labelled as the people who hate snow leopards. And even though it is the herders who live alongside the snow leopards, they don't have the agency to manage their interactions. When we plan the design of their corrals, we take away even that little agency from them to decide how they will keep their animals safe at night. Even benevolent conservation efforts can have limitations when implemented from the top.

Here was the difference between the Sumdoo herders and the herders in the Gobi. The Sumdoo herders had taken back their agency to manage their interactions by designing their own corrals, and this had led to a change in their attitude towards the snow leopard. In Mongolia, we, the experts, had planned and executed everything. While I had felt proud of the electrified corrals we built, the herders had not experienced the same pride. They were passive recipients of our charity.

Herders live an interdependent life with their environment in which the interactions between their livestock, their dogs and the wolves inhabiting the hills surrounding their camps – even the mountains themselves – all have a place. We didn't need to raise money to predator-proof the corrals. Instead, we should be giving the herders the agency they need to keep their entangled lives secure in a fast-changing world.

# PART III

# Mountains and Data

# 8

# Fortress of Gya

Gya is the second-highest mountain peak in the state of Himachal Pradesh. Although a towering 6,795-metre giant, it is as elusive as the snow leopard. Gya is located at the tri-junction of Himachal (Spiti), Ladakh and Tibet, and the peak is hidden so deep in a maze of other mountain peaks that it remained little known till the late 1980s. Approaching it from Spiti is extremely difficult. Gya sits at the head of a fortress carved out by the Lingti River. Lingti translated literally from Spitian means 'an instrument that cuts rocks'.

Among the many defences of Gya, the first is a row of fairly high but relatively easy peaks connected by high ridges: Kanamo (5,995 m), Chau Chau Kang Nilda (6,380 m), Tserip (5,890 m), Kawu (5,910 m) and Shijbang (c. 5,900 m). To enter this maze of mountains, one climbs over a couple of high passes before coming to a rugged valley with large grassy meadows. Then the mountaineer is faced with another ridge of three mountains: Geling (6,100 m), Runse (6,175 m) and Gyagar (6,400 m). The lowest point to pass through these peaks is at 5,900 metres. Finally one comes to the mighty Gya, a single monolith rock wall of 1,200 metres from a starting point of 5,500 metres. Early mountaineering expeditions in the 1990s were unsuccessful because they could not find the mountain. Then a couple of expeditions ended up on the wrong summit: some climbers realising as they stood on the summit, others discovering this when their summit photos were carefully analysed. In 1999 when a team finally made it to the summit, they found a rope fixed on the summit ridge and realised that a previous expedition in 1998 by the Indian Army had reached the real summit but thought they climbed the wrong peak. In 2010, for my dumpling research, I needed a study site which did not have any livestock grazing as a control for the experiment. I was also curious to see snow leopard habitats undisturbed by humans. When I mentioned to a colleague that I wanted to establish a site in the Lingti Valley, he said that with mountain peaks getting lost in that maze, how would I find the snow leopard?

Villagers in the foothills of the Lingti Valley told me that there are extensive plateaus inside the maze of peaks. Lush green pastures cover these plateaus from end to end. They described the ruin of 'Yulshikpo', an ancient village located inside the high valley, abandoned because of its remoteness. When I asked them about snow leopards, they told me that they had never seen one but the numbers of blue sheep were large. Some said that mountaineers had seen as many as 120 blue sheep in a single herd.

I was set on getting into the valley and I turned to mountaineering literature. Harish Kapadia – former president of the Himalayan Club and the leader of an expedition to the Lingti Valley in 1983 – had published detailed notes on routes and challenges. One route into the valley was to follow the Lingti River upstream, but the current was very strong and crossing the river was almost impossible. To take this route, we would have to wait until the river froze completely during the winter and walk over the sheet of ice. This ice sheet had started to build up but was still too thin to walk over. The only other choice was to climb over one of the ridges that flank the valley and climb down into it. Harish Kapadia recommended the Shilla Pass (5,670 m), the lowest point on the first line in the defence of Gya. We would have to ascend over 1,400 metres from its base and descend into the valley beyond in one push as there was nowhere to make camp. A difficult task but possible. I turned to the elders of the village for advice. They were unanimous in saying that November was too late to climb a pass that high, but they also agreed this was my only chance to see the Lingti Valley before next year. According to them it would be suicidal to try to walk over the ice along the semi-frozen Lingti River.

I was not in a position to wait out the winter and come back the following year. My PhD required me to finish a certain amount of fieldwork and writing by the end of each semester. I could have done the dumpling study without attempting to survey Lingti Valley, but the mountaineer in me drove me on. Winter climbing was for hardened climbers. The phrase 'it cannot be done in November' fired me up. This was what I had trained for. Since I was not deploying my training to climb eight-thousanders, I wanted to use it in my research.

I decided to climb into the Shilla Pass and avoid the river. I thought I saw a fail-safe measure in my plan. If winter weather arrived quickly and the pass became unmanageable, I could always walk back out of the valley over the frozen river. I needed a small but strong team to achieve all this: three or four climbers with everything they needed to survive in the high altitudes (rations, clothing, climbing equipment) carried on their backs. We would camp at the base of the pass on the first day. On the second, we would cross the Shilla Jot. For the next seven days we would survey the Lingti Valley and then return the same way. It was an ambitious plan; I was asking the team to carry loads of over 20 kilograms and climb mountains that needed mountaineering expertise to scale.

Five steps and again I stopped to catch my breath. Panting, I looked back. Sushil followed a few metres behind.

The two other members of our team were nowhere to be seen. 'Let's rest while they catch up,' I said to Sushil, and we sat down, worrying about the huge snow plumes we could now see rising from the summit and ridges of Kawu, the 6,000-metre peak above, a signal that the winds were picking up and the weather would soon deteriorate. Sushil and I turned our backs to the ferocious wind, our throats so dry that we did not speak. Some minutes later, we heard hard snow crumble under footsteps and turned to see that our teammates had finally arrived. White lips and snow-covered eyebrows; both of them dropped down next to us. I don't remember how long we sat in silence. Looking around, only the white glare reflecting from the snow met my eyes. Visibility had dropped drastically because of the drifting snow carried on the strong winds. A thin sheet of snow covered our backs, but no one thought about it. We had bigger things to worry about.

Hands cold, throats dry, exhausted; all the water we were carrying had frozen in our rucksacks. Dark storm clouds were building and within the next few minutes we would lose sunlight. With the wind picking up, the temperature plummeted to 30 degrees below zero. I checked my GPS altimeter. At 5,550 metres above sea level, we had almost arrived. The Shilla Pass stands at 5,670 metres above sea level and I could see the prayer flags planted in the pass snapping in the howling wind. Just 120 metres left to climb, and we'd cross over and descend into the Lingti Valley and the green upland pastures we'd come looking for.

The top of the pass promised an end to our misery. I had not expected such strong winds on the leeward slope. But I was not sure if we had it in us to reach the top. Our backpacks were heavy and we did not have the same strength that we had in the morning. The chill wind was sucking out every last bit of energy in us. These final metres were an even tougher climb and also much more exposed to the wind.

Sushil was the first to get up. The three of us silently followed. The wind had picked up and it was carrying a spindrift of snow with it. We could hardly see the man in front. The situation showed no sign of improvement. We would often be blown off balance and it kept getting worse. We had hardly climbed 20 metres when all of us sat down again. It was time to reconsider our position. Despite having already climbed 1,300 metres we were not in a state to climb these last hundred metres, not in this weather. And what lay on the other side was unclear. Even if we crossed the pass, would we find a warm place to camp? What if it was just as windy on the other side? The risk was too great but no one was ready to say it. After another long and silent break, Sushil said softly, 'It's time to turn back.' Everyone felt relieved; the toughest decision was made and I was grateful to Sushil for it. He was the eldest in the group;

these were his mountains; he knew how far was too far. But I was also disappointed and frustrated, because we could not do what we came for. After all the planning, effort and pain we had to turn back just 100 metres short of the top. We were turning away from the gateway to the hidden valley of Lingti. I knew I would try again; we had taken a beating but we hadn't yet lost hope. The mysteries of the Lingti Valley would have to wait a little longer.

Snow leopards epitomise the romance of wilderness, a solitary life in remote mountains with no humans around, a quintessential ideal made popular by Henry David Thoreau in his book *Walden* in 1854. 'I love to be alone. I never found the companion that was so companionable as solitude.' Writers of the nineteenth-century Romantic movement considered nature sacred and rejected society, industrialisation and cities as corrupt. The snow leopard has become a modern symbol for romantic ideals, yet the snow leopards are rarely ever truly away from people. Research on the snow leopard diet shows that a large proportion is comprised of livestock (up to 70 per cent in some places). Some biologists argue that livestock is essential for or subsidizes snow leopard conservation. In contrast, my research has shown that while snow leopards eat a fair amount of livestock, their populations increase only with the availability of wild herbivore prey. Örjan Johansson's research using satellite collars has shown that less than 10 per cent of protected areas in the high regions are large enough to have a viable population of fifteen or more snow leopards. And most of these protected areas are not free of humans. They contain a large number of villages that have been around for hundreds if not thousands of years. The parks, largely set up between the 1970s and the 2000s, were often designated as protected land without consideration of the existing human population and their basic rights. The pastoralists who lived in the high mountains were invisible to the administration.

Only a handful of snow leopard habitats are truly without any people. The Nanda Devi National Park in the state of Uttarakhand in India has 250 square kilometres of inner sanctum which is untouched except for an occasional mountaineering expedition.

Although this is one of the most coveted regions for climbers, the Indian government is restrictive in giving permission for mountaineers to visit. The rumour is that the government collaborated with the American CIA in the 1970s to set up a device on the summit of Nanda Devi, the highest peak in the region, to spy on Chinese nuclear and missile-testing work across the border in the north. The inner region is surrounded by high peaks and ridges, some of them reaching over 7,000 metres in altitude. The only access is along the deep Rishiganga gorge, with the towering Trisul (7,120 m) and Dunagiri (7,066 m) peaks on either side. The sides of this gorge are treacherous. Many expeditions attempting to climb mountains inside the inner Nanda Devi sanctuary find this section, where the entire expedition team and all the food and camping gear have to pass through the bottleneck, to be the hardest to negotiate. For an academic like me working for an NGO, getting permission from the government to be part of one such expedition or mount my own would be a difficult challenge. But it is more honest to say that I have never tried.

I had my eyes set on something even bigger. The Sarychat-Ertash Nature Reserve in the Tien Shan mountains of Kyrgyzstan is a contiguous stretch of 1,500 square kilometres of snow leopard habitat that has no other form of human use, at least not currently. A colleague described this as the place with the largest herds of argali in the world.

By early 2017, I had been the Director of the India programme of the Snow Leopard Trust for two years. The Trust had a small program in Kyrgyzstan. I had already set up multiple projects to monitor populations of blue sheep, ibex and argali in parts of India and Mongolia. In both of these countries, I had worked with the local herders and rangers and trained them to conduct surveys across large regions. I had many questions about Sarychat, such as how many snow leopards live there? What densities do ibex and argali achieve in the absence of competition from livestock? And what is the snow leopard's diet in a place where there is no livestock? Sarychat was my best bet to study the population density of snow leopards and their prey before humans arrived on the scene.

Organizing an expedition for snow leopard research is not very different from organizing a mountaineering expedition. You define your objectives and identify team members who have the skillsets required to achieve those objectives. Ideally, your team members should complement each other in their skills and get along well. You need all the right equipment and a way to get your team and the equipment to your base camp. Once the base camp is set, your team can start working. And of course, someone has to pay for all of this.

Kubanychbek Jumabay-Uulu, known to us as Kuban, the Director of the Kyrgyz programme of the Snow Leopard Trust, was already doing research in Sarychat-Ertash Nature Reserve. He had been deploying cameras to identify the snow leopards that live there and estimate their populations. He had a tendency to name them after Hollywood celebrities. What Kuban had not yet done was estimate the prey populations. In one of the few places in the world where wild herbivores did not have to compete with livestock, their numbers could indicate baseline population densities for a world without humans. I am not a misanthropist, and I believe that humans have as much right to live in the high mountains as any other species. My aim is to make conservation recommendations that allow for human and non-human species to live together with as little cost to each other as possible. But the scientist in me was curious to understand what the highest densities of wild herbivores and snow leopard would be if they were given a chance away from humans.

Kuban and the park administration were keen to have a scientifically robust estimate of the population of ibex and argali in the Sarychat-Ertash Nature Reserve and to understand how many snow leopards the prey could support, so that became the objective of our expedition.

I was also hoping to see a snow leopard. By now, I had seen them more than fifteen times, but this was not about racking up the tally mark – I would never leave Spiti if I wanted to simply increase my count. Seeing a snow leopard was the ultimate prize of working in the mountains. While there is always an element of chance, a thorough understanding of the mountains and its inhabitants – human and non-human – led to some of the most incredible snow leopard sightings. Such deep knowledge can only be gained by spending months, if not years, walking the ridges. Kuban had spent decades in Sarychat and I was hoping to learn from him.

Over the years, when the thought of working in Sarychat was brewing in my mind, I had discussed it with and excited two of my friends with the idea. Munib Khanyari and Prasenjeet Yadav were both younger than me and, like me, had won National Geographic Young Explorer grants at the start of their careers. We knew we could raise funds to support our research from the National Geographic Society if we banded together.

Munib was a Kashmiri from the beautiful city of Srinagar. His parents moved to Mumbai in the early 1990s, at the peak of the violent militant extremism in Kashmir, in order to provide better educational opportunities for Munib and his two siblings. Munib was thin and tall with a pointy beard and a long face. He was un-apologetic about how much he talked and how much he loved to walk in the mountains. He was a joint PhD student with Bristol University and NCF and I was part of his advisory committee. If there was one person who matched my enthusiasm for mountains and the animals that live there, it was Munib.

Prasen grew up in Nagpur, a city in Central India, and had spent a lot of time on his family's farm in the outskirts of the city, sur-rounded by dense teak forest filled with tigers and leopards. He had joined NCBS as a PhD student studying tigers when I was near the end of my PhD on snow leopards. We hit it off immediately. One day he told me that he wanted to quit his PhD and take up his passion: wildlife photography. Soon he had assignments from the *National Geographic* and the BBC. Prasen and Munib, along with Kuban, were a perfect team for such an expedition.

The Double Observer Survey method requires researchers to work in pairs. Prasen was going to document the expedition, so I needed one more person for our team. Suraiya Luecke was a young Indian-American undergrad student interning with Kuban in Kyrgyzstan. She had impressed us with her work in the office and she was keen to spend time in the mountains. Suraiya was tall and strong with long curly hair. She had already been hiking in the mountains around Bishkek and she was ready for the big mountains.

When Prasen, Munib and I landed in Bishkek, the full team met for the first time. We were gathered in Kuban's office with camera-trap photos of snow leopards named Brad Pitt, Jennifer Lopez, Tom Cruise, Angelina Jolie and others covering the four walls. These are all from Sarychat, Kuban explained.

'Is Brad Pitt as handsome and Tom Cruise as adventurous as their Hollywood counterparts?' I asked.

'Yes, and Angelina and Jennifer are just as beautiful,' Kuban responded. 'It makes it a bit awkward to talk about their reproductive behaviour though,' he added with a bit of embarrassment.

A small map of Kyrgyzstan and a big map of Sarychat-Ertash Nature Reserve were open on the table. While I worked out how we could survey the entire region, Kuban figured out a way to keep the team supplied with food and gear. I punctuated my monologue with tips for Prasen on capturing the key moments of the expedition and important methodological assumptions for Suraiya to get the hang of the Double Observer Surveys.

At the end of the meeting, Kuban pointed out that Munib, with his Kashmiri features – fair skin, long face, thick brows, moustache and beard, thin tall body with long limbs – looked like the Uigur people in Kyrgyzstan. A week later we were to learn that this similarity could have dangerous consequences in the place where we were going.

During our briefing in Bishkek, we finalised a two-fold objective. As well as estimating the population of argali and ibex and snow leopards in the nature reserve, we would also research populations in the adjacent hunting reserve. The Koilu hunting concession was used for cattle grazing as well as grounds for sports hunters to shoot big trophies of ibex and argali. In 2017, a single licence to shoot an ibex cost upwards of $2,000 USD and an argali over $20,000 USD. A large proportion of the Kyrgyz countryside is parcelled out into hunting reserves and reserve managers pay a fee to the government to manage them and earn revenue by hosting hunters. If you pay enough, you could get a licence to shoot everything from partridges to mountain ungulates, just short of a snow leopard. In theory, hunters paid the licence fees to the government and then paid a hunting outfit to rent a lodge, cars, cooks and guides to hunt a male ibex or argali above a certain age determined by their horn size.

Collecting animal trophies may be as old as humans, but this is a poor justification for the current practice. Modern-day trophy hunting is different from other forms of hunting because the primary motivation is to acquire a trophy and in the process experience the thrill of killing a magnificent animal. When hunting for food, the hunter is inherently concerned about the sustainability of his food resource, whereas the trophy hunter puts a premium on the killing of the rare species. The rarer the species, the higher the demand, and the higher the licence fees. This leads to an incentive to keep a species rare because it brings in greater fees.

Kuban's concern was that the trophy-hunting industry was not well managed and malpractices were affecting the populations of ibex and argali, not just in these hunting concessions but also inside protected areas like Sarychat. He knew that sometimes when the hunters were not happy with the size of their trophy, they bribed the guides to bury it and hunt a new trophy. Hunting reserves on the edge of Sarychat commanded top dollar because they boasted higher numbers of ibex and argali spilling over from the reserve. Kuban showed us YouTube videos of hunters violating park boundaries and shooting inside the reserve. He had also photographed hunters inside Sarychat on the camera traps that he had deployed for snow leopards.

Licences for trophy hunting are an important source of revenue, especially foreign exchange, for several snow leopard range countries, but the practice comes with controversy and debate. A single licence to hunt a markhor goat in Pakistan costs more than $100,000 USD. This programme is often cited as a success story as a large share of the licence money is shared with the local community for their development. However, critics allege that hunting incentivizes illegal persecution of snow leopards and wolves because the carnivores deplete the population of markhor, including the ones with large horns, which make up a part of their diets. An average snow leopard kills nearly fifty large herbivore prey in a year. This number is bound to include a few trophy-sized males, which means the snow leopard can become an enemy of a trophy-hunting programme that's financially incentivized to protect markhor for their clients. Occasionally, we heard reports of Russian oligarchs hunting argali from helicopters with high-powered rifles. A photo circulated among our team of an unidentified hunter posing with a snow leopard that he has just shot. The year and location are unconfirmed, but it is most likely to have been taken in the Russian Altai mountains soon after the break up of the Soviet Union.

In the 2010s an ignominious self-identified researcher had secured permission from the Mongolian government to 'put down' four snow leopards to study their reproductive biology. The perverse idea was to have wealthy hunters bid for the right to hunt these four poor snow leopards. The global scientific and conservation community was unanimous in pressuring the Mongolian government to withdraw the permission before the idea could go far.

The snow leopard is the only species of big cat that has never been the object of trophy hunting. Their small size and lack of aggression towards humans has always made them an unworthy opponent to those hunters who wanted to pit themselves against the might of the Big Five of Africa, the Indian tiger or the impressively horned Himalayan mountain goats and sheep.

Studying today's trophy hunting and its effects on snow leopard and wildlife conservation in central Asia is difficult because of the opaqueness with which most countries operate. The large amounts of money involved leads to bribes and secrecy. Kuban's more modest goal of demonstrating the difference in the population density and demography of ibex and argali inside a nature reserve and a hunting concession seemed more achievable.

Sarychat is in the remote north-eastern part of Kyrgyzstan, bordered by China on the eastern side and Kazakhstan on the north. The Koiluu Hunting Concession abuts the reserve on the north-east side, snugly under the Chinese border.

We visited the bazaar in Bishkek to stock up before driving to Sarychat. Kuban pointed out the stores run by Kyrgyz, Russians and Uigur people. He knew exactly what to buy from which store. I wanted to keep my mind free to think about planning the survey in the field so I wandered around without much purpose. Prasen was having a great time taking pictures. Munib and Suraiya had a long list written by Kuban and were busy shopping. The last item on the list was bread. We visited a local bakery at night. Kuban had already placed an order for seventy-five loaves of the round pizza-like soft Kyrgyz bread. The top was warm golden with a glaze of butter and each loaf carried the bakery's signature design, a round necklace of diamonds. I could smell them a mile away. The boot of Kuban's 4×4 Toyota was half-full of equipment: tents, sleeping bags, spotting scopes and binoculars, heavy snow boots, a box of jams and spreads, lots of cheese, canned meat and vegetables. The bread took up the rest of the space, all the way to the roof of the car. In India, Thinley or Sushil would have been unhappy about packing the bread next to the boots. Different cultures did things differently, even on expeditions. I was quieter than usual because I was thinking through the scenarios we faced. We started for Sarychat early the next morning for our full-day drive. As we approached Lake Issyk Kul, Kuban announced that he was friends with a local falconer who lived close by and wanted to know if we were interested in meeting him and his birds. I was shaken out of my expedition zone and felt as excited as a child. Before the others had begun to contemplate the proposition I had agreed. Someone tried to ask if we would be delayed in reaching our destination. I interjected that this might be the only opportunity to meet a falconer from this part of the world. That settled it.

We drove up to a house on the outskirts of a small Soviet-looking town. The lake was visible in the distance and the water extended to the horizon. As I stepped out of the car, I could hear a golden eagle. Then I saw one, sitting on a small stool, its hood on. Next to the eagle was a sleek, sinewy dog. I recognised him as a taigan, a sight hound used for hunting in Central Asia. A middle-aged man emerged from the low wooden door. His face was tanned, with lines along his forehead showing the endless hours he had spent outdoors. He wore a woollen cap and walked with confidence and purpose. Kuban introduced him as one of the best falconers in Kyrgyzstan.

His taigan dogs were friendly; then he showed us his birds. Two golden eagles, a juvenile and an adult female; a goshawk; a peregrine falcon; and a hobby. He had had a saker falcon which had been released only a few weeks ago. The falcons reminded me of the stories of Amur falcon migrations that Bhagya and I had shared during our courtship.

The juvenile golden eagle was almost black and called continuously, her golden beak opening wide, her shrill voice piercing my skull. Her black talons contrasted with her yellow feet and were almost as long as my index finger. She looked us right in the eye, like an angry human. 'Don't go too close to her; she can attack your face and pull out an eyeball,' the falconer warned my colleagues. I asked him which bird was his favourite.

'As a hunter, it will always be the goshawk. But tourists pay more money for the golden eagles.'

'What do you mean? Is it more exciting to hunt with a goshawk than a golden eagle?' I asked, surprised.

'Golden eagles need a lot of food, so you have to provide a lot or be out there hunting all the time with them. They are great at hunting foxes, but there are few foxes these days. You can hunt hares with them, but you can also hunt a hare with a goshawk. A goshawk needs a lot less to eat than a golden eagle and they are more exciting to hunt with.'

'What about the peregrine falcon? I'm always mesmerized by the peregrine falcon when I see them in the wild.' The female that was sitting on my fist was smaller and more docile than I had imagined.

'Peregrines lack focus,' he said in a resigned tone. 'When you release the bird, she wants to fly around and explore the whole place. Even after she hunts, she will wander around for some time before coming back to you. Sometimes I have to coax her for hours before she returns to me. They are good at flying very high and then diving at a flock of pigeons but it is easier to hunt pigeons with a gun. They are sitting on every roof in town.'

'The peregrines won't bond with you the way a goshawk does,' he said, looking at the steel-grey goshawk with rusty streaks on her belly perched calmly on his leather-clad fist. In contrast, the peregrine on my hand had a black helmet and a streamlined dark-grey body shaped like a dart and looked around anxiously. Its smaller cousin, the Eurasian hobby, had a similar helmet but orange streaks on her belly and sat on a wooden perch a few feet from us. Prasen was busy taking pictures while Munib and Suraiya were petting the dogs and the golden eagles. Kuban helped translate my English and the falconer's Russian.

'So you hunt foxes with the eagles and pigeons with the peregrines. What do you hunt with the goshawk?'

'Everything,' he said with a smile on his face. I could see that he loved his goshawk and was happy that someone was interested in birds other than the eagles. 'Hares, partridges, quails, larks: anything that sits on the ground or perches in a tree, the goshawk will catch it.'

We spoke about how he trains the birds. He mentioned that he had caught the hobby from her nest earlier that spring. He said he would teach her and hunt with her for a few years before releasing her.

'Like you released the saker falcon?' I asked. It turned out that he had caught the saker falcon from her roost and not her nest. Birds taken from the nest are very young, only a few weeks old, and they are dependent on the handler and bond with them naturally. Those caught at their night roost are adults and are trained differently. They are free birds who have to be taught to become dependent on the falconer. Then Kuban mentioned something interesting. He said his grandfather was a renowned falconer under Soviet rule and the first person in the world to breed saker falcons in captivity, for which he had been decorated with a civilian award.

'But they are having a tough time today,' Kuban added. He had witnessed the customs department confiscate thirty saker falcons that were being shipped illegally to the Middle East. I could see the falconer's face fall. He was suddenly ashamed of being a falconer.

'Falconry is a traditional art, livelihood and lifestyle. The birds are your family. You don't trade in them,' he said with a sad face.

'When I release a bird I feel sad and I feel happy. It's like my child growing up and going out into the world. A traditional falconer carries a stigma if their bird dies. Other falconers will look down upon you.

'But now, new people are entering falconry. They think that golden eagles will help them make money from the tourists. They try to sell birds, and there is a large illegal demand for falcons in the Middle East. But it is a lot of work, and they are not prepared for it.'

King Abdullah of the United Arab Emirates is on a mission to revive falconry and to develop its global popularity. Dubai is organizing all kinds of falconry competitions. They have set up a world-class hospital for falcons. Falcon racing has been introduced there, and the prize money is comparable to that of the leading horse races around the world. UAE has set up one of the best facilities for falcon breeding in order to stop the capture of birds from the wild. UNESCO lists falconry on the Representative List of the Intangible Cultural Heritage of Humanity. The argument is that falcons can be saved by the art of falconry – the same argument being promoted for trophy hunting. The long-term impacts of falconry on the populations of birds in the wild are uncertain, but hundreds, if not thousands of falcons are captured and traded illegally around the world every year.

India too has a long tradition of falconry and hunting, called shikar. The portraits of kings and leaders often featured their falcons. In almost every painting of Guru Gobind Singh, the first spiritual and military leader of the Sikhs, he is pictured with his goshawk. Paintings of the Mughal kings similarly include falcons and hunts of tigers and lions. Chand Bibi – the warrior queen of the Deccan, the place where I was born – is painted with her horse galloping and her falcon on her arm, attended by a small group of armed women. Hunting and falconry blended with European and British traditions during colonial rule.

However, when compared with the United Arab Emirates, India chose a diagonally opposite route to conservation. After the decline of India's tigers, which went from a population of over 40,000 in the nineteenth century to a low of 2,000 by 1970, the Indian Parliament passed the Wild Life (Protection) Act in 1972, banning hunting in all forms barring a handful of tribal practices. Trophy hunting by colonial officers, rajas and the aristocrats had been one of the primary causes of the tigers' near extinction. The new law acknowledged that without a blanket ban on hunting there was no hope for the tiger or any other species of wildlife in India. The law also banned all forms of falconry and owning, possessing and trading of native wildlife species. It ended centuries of tradition and culture, but created room for wildlife to coexist with people in a different way.

For me, meeting the falconer and seeing his birds on the banks of Lake Issyk Kul was a small tryst with my own history. Today, there is sporadic demand for snow leopard cubs from private menageries. It is not surprising that some people want to keep the 'cute and cuddly' snow leopards as pets. Legal and illegal trophy hunting and falconry creates the spaces within which the trade and ownership of snow leopards is made possible and even legitimized. So far, the demands are not a big threat to snow leopards in the wild but it is difficult to predict the future.

The drive from Lake Issyk Kul to Sarychat went past abandoned old Soviet buildings. Lake Issyk Kul, one of the deepest in Central Asia, had been a torpedo-testing facility under the Soviets. 'During the Soviet times' was the opening line of every story Kuban told us. It was late, dark and cold by the time we reached a camp at the edge of Sarychat-Ertash Nature Reserve.

We had planned to survey the park over the next seven days. The car would go for another few kilometres to the next camp but travelling thereafter would be on horseback. We were joined by three local rangers, – Mirbek, Temirbek and Askat – and eight horses. I had grown up riding horses in my village, but for Prasen, who was less experienced with horses and who was also cataloguing our research through photography, riding was going to be a challenge. Suraiya had ridden a horse only once in her life and 'that did not end well', she told us. But she gritted her teeth and was ready for whatever lay ahead. Munib would have preferred to walk, but this was not feasible considering the equipment and food we were carrying, and keeping pace with the horses would not be possible. After some discussion, it was decided that I would ride a horse named Shaitan (the Devil) and as long as I kept him under control, Kuban and the rangers were free to pair up with Suraiya, Munib and Prasen to help them on the journey. Each day we would split into three teams of two riders to survey a designated valley and then regroup at the end of the day at a new campsite. Either Kuban or one of the rangers would be part of each team because they knew where it was safe to cross rivers and streams, and the locations of the campsites. The other person in the team would be Munib, Suraiya or me, while Prasen did the photography and filming.

On the first day, we rode uphill. It was autumn, and the mountains had turned golden, but they shone bright yellow in the early morning light. The higher ridges were covered in fresh snow. Shaitan walked and trotted steadily and I was developing a friendship with him. My life would depend on him in the coming days. Today, we had to cross a high pass and enter the Sarychat River valley. The climb to the pass was steady but demanding on the horses. We stopped often to scan the surroundings with binoculars to find argali and ibex. At first, everyone struggled to hold their binoculars steady while mounted, but this would soon change as we became more familiar with the horses. As we climbed into the pass, the temperatures reached sub-zero, and I needed gloves to hold the reins without hurting my fingers. Beneath Shaitan's hooves was a dried autumnal meadow with withered flowers. I had spotted a herd of argali in the distance, but I could not see any details because Shaitan would not stand still; he was excited to descend from the pass and get to the next camp which he knew very well. Perhaps he was trying to get out of the dangerous places and into the safety of the camp. I would need to gain his trust for these surveys to work.

I dismounted and held the reins tights. The last thing I wanted was for Shaitan to gallop to the camp while I had to walk the rest of the way, holding my head down in shame. I found a boulder and sat down, steadied my hands and started counting the argali. As my binoculars focused, I realised that an incredible picture was emerging in front of me. A large herd of over a hundred argali were gathered, and nearly a third of them were large males. This single herd contained nearly as many argali as the entire population of the species in India, or the entire population of Tost-Tosonbumba Nature Reserve in Mongolia. About fifty large males with horns measuring more than a metre, weighing over a hundred kilograms and standing taller than 6 feet, walked slowly in single file up the mountainside. They looked like a river flowing upstream. My hands were shaking from the excitement. I was making notes in my head on the age and sex of the animals as they went past a boulder in ones, twos and threes.

When they were beyond view, I brought down the binoculars and made my notes in my small pocket diary. I looked over towards Shaitan, who was grazing peacefully next to the rock. To be on a mountaintop with my horse grazing next to me while I watched wild argali, with likely a snow leopard nearby – this was the greatest reward of my work.

We reached the camp well after dark. I took Shaitan to the stable area, and one of the rangers helped tie him and gave him hay and feed for the night. We were expecting another slow, long day ahead.

We ate a supper of bread and cheese outside, on a boulder. The Milky Way hung overhead like a chandelier. If you stared hard at it, and when your eyes had softened to the darkness, you could perceive its depth and sense that some stars are closer than others. Starlight illuminated Khan Tengri, standing taller than 7,000 metres in the distance. I could hear a river gurgling nearby in the dark. Kuban assured me that we were high enough that even if there were a flood at night, the water would not reach us.

Kuban had a bath in a small stream the next morning while the temperatures were still sub-zero. The rest of us watched from a distance, struggling to brush our teeth in the cold. He came back and announced that we needed to leave early because we had to cross the river, and the water would be high later in the day when the snow and glaciers started melting.

The water was freezing and the river was over 50 metres wide. Kuban had picked a spot where the water was not too fast to sweep our horses away, and not too slow where it would be too deep. Our horses would wade across with us on their backs. The water was expected to reach the horse's belly, maybe a bit higher. Kuban made it sound easy and took the helm like a true leader. We followed one by one. One of the rangers was to bring up the rear and I was just before him, keeping an eye on everything that was going on ahead. Prasen's horse stumbled and I could see him tense up. He was more worried about his camera equipment than his own well-being. Suraiya, who was riding a horse only for the third time in her life, kept calm and followed one of the rangers, who had connected a rope between his horse and Suraiya's. Shaitan walked like a boss. Once everyone was safely across to the other bank, Kuban and I shared a quick look. Our colleagues did not know what a difficult crossing they had achieved.

That night over the campfire, Kuban told the story of the day when his horse slipped and he fell during the same river crossing many years ago. At first, Kuban tried to swim, but the current was too strong and his jacket became heavy with water. He could see his colleagues galloping along the bank to keep an eye on him as he flowed with the current. 'It is clichéd to say this, but the faces of my family members flashed in front of my eyes as I started going under and bobbing back out,' Kuban said. He would black out, and the cold water would wake him up again. Soon, he was too numb to feel the cold. He lost sight of his colleagues. He thought he was already dead, and these were visions from beyond. Then everything went black.

His colleagues had been following him. They knew that his only chance was to grab the willow trees that hung low over the river a kilometre down from where he had slipped, but they were not sure if he would be conscious or even alive for that long. If he missed it, then there were dangerous rapids ahead and little chance of surviving. Kuban was lucky to get tangled in the willow branches, with his face towards the sky and not the water. Even in his unconscious state, he grabbed the branches with his freezing hands and hung on. His colleague had to pry open his grip to get him out.

'I am glad you did not tell this story before our crossing. I would not have gotten into the water if I knew how bad it could get,' Prasen said.

'We have to cross back in three days,' Kuban replied. 'And there is no other way out but this.' He smiled.

Sarychat is a jewel of natural beauty and a paradise for wildlife. Large herds of argali strutted past our horses like scenes from science-fiction movies. The big males were almost as tall as our horses. They were concentrated around the base of the valley while massive herds of ibex clung to the walls of the mountains. The animals live in a vertical world. We were surrounded by high snowy peaks; the valley was covered in a mat of golden grass and the ground was soft dirt, perfect for riding horses. Snow leopards lurked in the rocky cliffs that stood between rolling hills and the big mountain peaks. Munib and I joked that we were seeing more argali and ibex in this short visit in Sarychat than we would see throughout the year in our travels across the rest of Asia.

Shaitan became nervous a few times and I wondered if he were smelling or sensing snow leopards. One of the rangers was of the opinion that the horses felt the presence of wolves more easily than snow leopards. On more than one occasion his horse had refused to move and he had then spotted a pack of wolves in the distance. But I was conscious that his observations included bias. While it is possible to spot wolves in the distance, it is almost impossible to see snow leopards as they are so well camouflaged. It is likely that on some of the numerous occasions when the horses refused to walk on and we did not see anything in the distance, that a snow leopard was hiding nearby in the ravine or the crags; not with the intention of attacking us but just to get away.

Kuban took us to check one of his cameras and we found that it had taken a picture of a Eurasian lynx two days earlier and a snow leopard two days before that. They were around us but we could not see them. We were a largish party of eight horses with a lot of equipment, and making a lot of noise, enough to warn them early of our presence.

The next day, during the survey, Shaitan refused to walk on in a meadow. No form of coaxing worked. I scanned the slopes in front of me with my binoculars for wolves and snow leopards. When I was convinced that there was nothing to see, I hopped off and led

Shaitan with his reins. The reason for his hesitation became apparent only a few metres later. A dead ibex was lying near a boulder, only skin, bones and horns. The flesh had been plucked clean, clean enough for me to guess that it was a snow leopard's work. The ground was too hard for pug marks and yet I was certain that the kill was a snow leopard's because the bones were still connected with each other and the entire skeleton was intact. A pack of wolves would have ripped it apart and taken chunks. The thought of hiding near the carcass to see if the snow leopard returned at night crossed my mind but it was surrounded by fox poop. A snow leopard would not allow a fox to spend so much time near its kill. It was evident that the snow leopard had moved on.

We know from radio-collaring studies that snow leopards kill a large prey like a blue sheep or an ibex every seven to ten days. The carcass in front of me was fresh, which meant the snow leopard was unlikely to hunt again in this valley over the next few days. The snow leopard would be less active during this time after a feed, so the chances of seeing this particular individual were low.

After completing our surveys in Sarychat, we bid farewell to the rangers and headed to Koiluu Hunting Concession. I had grown fond of Shaitan and I was sad to leave.

The landscape in Koiluu was similar to that in Sarychat. The valley was wide and lined with golden grasslands, with rocky cliffs on the sides and snow-covered ridges. But the grasslands were dotted with thousands of cattle. At the entrance to the valley was a hunter's lodge. It was plush with central heating, solar electricity and satellite internet. It was so alien to its surroundings that it might as well have been on the moon. We camped as far away from this decadence as possible.

It took us a minute to notice the three sets of ibex heads neatly arranged on a mat in the grounds of the lodge being readied for mounting and packing. The scimitar horns poked out of buckets of steaming-hot water. Our mood became morose. We drove into the valley and met the two hunters who were decked out in expensive outdoor clothing. We exchanged pleasantries and discussed ibex numbers in the valley. We agreed that the numbers were low.

They said they were happy with their trophies but were critical of the management without being explicit about it. There was not enough trust between us to talk about the sensitive subject of trophy hunting.

Much of Koiluu Valley is connected with roads so we did not need horses, and we planned to walk the few valleys that did not have roads. Munib was very happy to be back on his feet. At breakfast on the first day, Kuban started telling us stories of the Uigur people escaping the Chinese regime on the other side of the international border less than 30 kilometres away. The Uigur people are from the Xinjiang province of China which abuts Kyrgyzstan. They are fending off Chinese pressure to integrate so that they can save their unique culture and way of life. Global tensions reached a crescendo when the West put sanctions on goods produced in Chinese factories, alleging poor treatment of Uigur workers. A large number of Uigurs who were protesting were either imprisoned or sent to 'correction facilities'. Many tried to escape.

Kuban told a particularly grim story about fourteen Uigur men who had crossed the border into Kyrgyzstan and were surrounded by the Kyrgyz Army. The army wanted them to surrender but they feared deportation. The scuffle became violent and all fourteen people perished. Kuban explained that the Kyrgyz Government finds itself in a difficult bind. They cannot afford to antagonize their powerful neighbours by providing asylum to Uigur people in Kyrgyzstan, nor can they afford to deal with the international pressure for not doing so. The government had adopted the tactic of repelling as many immigrants as possible without recording it on paper.

That is when Kuban looked straight into Munib's eyes and said that if locals mistook him for an Uigur immigrant he should not try to run or be seen to be getting angry. Kuban advised him to stand his ground, keep his hands where everyone could see them and repeat 'Kubanychbek, *irbis*' meaning 'Kuban, snow leopard'.

'They will know you are with me.'

Just as Kuban had predicted, when Munib was walking in a valley counting ibex, he heard a whistle, and a rider galloped towards him with a rifle on his back and a hunting dog by his side.

Munib climbed a boulder to stay away from the dog but sat down with his hands over his head. When his pursuer heard the words 'Kubanychbek, *irbis*', he was disappointed at first but then invited Munib for a cup of tea. Munib politely refused and continued his surveys for the ibex.

The differences between Sarychat, a nature reserve, and Koiluu, a hunting concession, became clear after our thorough survey. Sarychat had close to a thousand argali, while Koiluu had none. The argali habitat of Koiluu was overrun with cattle. The density of ibex in Sarychat was four times higher than in Koiluu, though in Koiluu the ibex stayed together in larger herds – most likely due to the higher predation pressure from hunters. Cumulatively, the herbivore densities in Sarychat were three to eight times higher than most places I had worked in the western Himalaya, the Altai mountains of the Gobi and the Tian Shan mountains. Only in the Lingti Valley had we seen similar densities of blue sheep.

There was enough wild prey and enough land in Sarychat to support about twenty-five snow leopards. These might be among the only snow leopards who don't have to make the daily choice between hunting domestic livestock and wild herbivores.

Six years earlier, a day after the team's failed attempt to cross the Shilla Pass to reach the Lingti Valley, we hiked to the village of Lalung, which sits right on the banks of the river as it exits the sanctum of the valley between the steep and tight cliffs. It looks here as if the river is emerging from a cave. I tried to find people who had been inside the valley before. Many claimed to have journeyed there, but very few people could give me straight answers to my questions. Finally, we met Thukten. We called him Lalung Thukten for we had another Thukten from Kibber in the team.

Lalung Thukten spoke straight. He told us many things about the valley and how to make it in and, more importantly, how to get back out. I asked him about snow leopards and he said there were plenty inside. I asked him if he had seen any, and he gave me a puzzled look. Snow leopards are not something you see, he wanted to say. They are something you sense and know.

'There will be hundreds of signs all the way in and out,' he said at last, sounding a little offended.

The best route into the fortress was over the frozen river but there would be sections where we might have to get our legs wet.

I was the only person in our team who knew how to swim and even then the water in the river was too fast, the cascades too dangerous and the temperatures too cold.

'It will be like walking on a white carpet,' Lalung Thukten joked.

This sounded very dangerous. We did not have crampons to put on our boots. One slip over the ice, and we could be swept off by the current of the river below. Even if we survived the fall, hypothermia would kill us in a matter of hours.

I sat down with everyone in Lalung Thukten's house to pick a small team of people who would make a last-ditch attempt to walk over an iced-up river and cliffs without the right gear. Despite Rajender's warning against last-ditch attempts, I could sense a certain confidence in Lalung Thukten's voice which convinced me that we could do this. There was a small window of opportunity and I was determined to try.

To my surprise, everyone wanted in. A form of brotherhood had formed in the team and nobody wanted to be left behind. Sushil advised me that having more people on the team was good. Once we had crossed the tricky sections, we could set up a base camp where one person would be responsible for keeping everyone well fed and cheerful while the others went out and surveyed the place in teams of two each. 'We could divide the work between us and be done sooner,' Sushil said. I agreed. Everyone had volunteered but I was responsible for the team's safety and I was committed to it. If we were faced with tough situations, then this time it would be me making the decision. Until we set up our base camp, I had to think like a mountaineer.

During my first winter in Spiti, KP and the teacher went back to Kibber to party on New Year's Eve. I was not in a party mood so I stayed back and Takpa came to spend the evening with me, feeling sorry for my loneliness. That night, Takpa and I spoke about Kanamo, the mountain that stood just five metres short of 6,000 metres and overshadowed Kibber. I told him about my experience speed-climbing it solo during my summer internship. I had taken less than five hours to go up and eight hours for the whole thing. Takpa said that he had lived in that mountain's shadow his whole life and never made it to the top, and just like that, the two of us decided to climb the mountain the next day to celebrate the New Year's sunrise.

Kanamo does not require any technical climbing expertise, but it was cold. When we left Takpa's house, it was −19° C at 4,300 metres. We were quick and in less than seven hours we reached the exposed summit at 5,995 metres. Our hands were frozen stiff. We took the customary summit photo and rushed back down. The cold had penetrated our bones and we felt as if we had escaped with our lives. When we reached Takpa's house, his wife asked him matter of factly if he made it all the way to the summit as 'after all it was your third attempt'. I was hoping that this experience of winter climbing had taught me enough to succeed in my attempts to survey Lingti while walking over a frozen river.

We left the safety of Lalung village and walked upstream along the river between the two large cliffs that stood like sentinels over the Lingti Valley. There were ten of us, and the group was excited to explore a place that nobody had visited in decades. As we entered the cliffs, the air became colder and greyer. The low winter sun never reached the river here. The plan was to walk along this section for three days and set up a base camp at a place called Phiphuk. At Phiphuk, the valley opens up, and we would be able to reach the plateaus, which had pastures and hence herbivores and, hopefully, snow leopards.

The first ice crossing was easy: a bridge of sheet ice across the river. We skated across on our hiking boots. Then came a crossing where the middle section had fallen in because the water levels had dropped. We would have to jump from one ice shelf to the other. Lalung Thukten brought a bag full of pebbles and dirt and made a runway for everyone to get some traction. He threw some dirt and pebbles onto the other shelf where we were likely to land. If we missed, the fall was over 15 feet into a narrow passage of water below. Lalung Thukten was the first to jump. I tied a rope around his waist, made a firm anchor by the side of the river and got ready to belay. Everyone watched anxiously. We would all have to do it, and it was time to learn from the master. Lalung Thukten jumped without thinking twice. There was relief on everyone's faces. Once half the team was across we hauled over the bags and the equipment using ropes. I was the last to jump, roped and belayed by Thinley, who was doing it for the first time.

We would do this drill about twenty times during the next ten days of the expedition.

On the third night, I lay awake in my sleeping bag next to a small fire of dying embers. The constellation Orion was overhead between the two steep cliffs that flanked me. The Lingti River was only a few feet away, completely frozen as this part of the valley received no sun in the winter months. I sat up to stoke the fire when two distant eyeshines caught my attention. My heart skipped a beat. Could it be a snow leopard? I tried hard to see but could not make anything out. Eventually, the animal moved closer. From its pace and movement it seemed small, but when it came within the range of my torchlight, I was surprised to see a stone marten. I had never seen one of these carnivorous weasels in my years of wandering the Himalaya. Ten years later, I would find out that stone martens are quite common but they are completely nocturnal in the Himalaya, making them one of the hardest species to see.

When I woke the next morning, I was in for a bigger surprise. I could see the pug marks of a snow leopard that had walked up to us along the frozen river and switched to the other bank only 10 metres from where we lay asleep. The wide paws made light impressions on the hard snow. In stretches with soft snow, there were gentle brushes where its tail had dragged behind. This was the first sign of snow leopards in this part of the valley.

We reached Phiphuk three days later. This would be our base camp, our home, for the next week to ten days. The area was a small grove of willow on the bank of the river with an ice bridge to cross over. The sun never reached this spot, but it was well pro-tected from the wind. The absence of wind kept the spot relatively warm, even at night. There was no risk of rocks falling on us from the surrounding cliffs, but I did not like the long slope that lay to our west. If heavy snow fell, then we were directly in the line of an avalanche. Lalung Thukten assured me that it never snows more than a foot in Lingti, but I remained worried for the rest of the time we camped there.

Every morning, four teams of two people would pack some food and head towards an area that I would mark on the map. Once we reached our spot, we would identify animal trails to look for signs of snow leopards. We would also look for herbivores. Everything would be marked with a GPS. If we found fresh scats of the snow leopard, the teams were equipped with a small kit to collect a sample for genetic analysis. After the first day of surveying, the teams returned to Phiphuk and we became lively with stories. Our base camp manager made us potato dumplings and a chutney of wild allium (onions) that he had foraged around the camp. Everyone had seen a lot of blue sheep. Snow leopard signs were all around. Looking at the habitat and the numbers of blue sheep, the team were confident that this was snow leopard heaven.

We surveyed every day but as our time in Lingti started to wind up, I was doubly anxious – about not having seen a snow leopard, and about returning through the narrow passage, like crawling under the mountain. Rajender's voice echoed in my ears that 'more accidents happen on the way back down from the mountain'. I remained watchful throughout our retreat back to Lalung.

The expedition was a successful failure. We had breached the fortress of Gya, surveyed the Lingti Valley and came back home safely. I was elated at the number of snow leopard signs and the number of blue sheep we had seen. But we had failed to conduct the Double Observer Survey to estimate the population of blue sheep, or to collect enough scat samples to estimate snow leopard numbers. We had lost precious time toiling around the edges of the Lingti Valley in order to find a way in and had worked day and night to get a glimpse of the wildlife in this remote part of the Himalaya. I had not yet been able to answer the questions about the snow leopard's ecology in a place where there were no humans. This was like climbing the hardest section of the mountain but not reaching the summit. It doubled my resolve to come back and try again.

Lingti became my obsession for the next few years, and I'd return season after season until I had found my answers.

The next summer, I returned with a smaller team and a gang of yaks. We rode the yaks across the torrent of the river, and this time we were able to estimate the blue sheep populations in the valley. The following autumn, I returned with better climbing gear and we were able to estimate the snow leopard population after collecting enough scat samples and running genetic analysis in a laboratory in Bangalore. We found eight snow leopards in an area that was half the size of Kibber.

Our results showed that Lingti had higher densities of snow leopards and blue sheep than anywhere in the western Himalaya. Snow leopards in the Lingti Valley fed almost exclusively on blue sheep. They rarely ate hares or partridge even when they were abundant. There was no livestock to supplement their diets. The fate of snow leopards in Lingti was tightly linked to the fate of blue sheep.

# 9

# Keystone or Capstone?

During my master's work studying the blue sheep in 2007–08, I spent a lot of time simply observing them going about their day. I would wake before sunrise, have a cup of tea, check the thermometer and head out to look for my blue sheep herd. I fancied myself a shepherd with an ice axe. The temperature remained below –10° C throughout the winter.

I would usually find the blue sheep where I left them bedding down the previous day. Once found, I would note what they ate, how often, what they did, and how frequently. In the end, I spent so much time watching this particular herd of sixty blue sheep that they grew used to my presence and allowed me to approach within 20 feet. Proud of their trust, I had to restrain myself from becoming protective.

On winter mornings they would start feeding at first light and continue till midday when they would rest, ruminate and chew their cud. This was my opportunity to eat my cold packed lunch. It often included a bar of chocolate, some biscuits, *tsampa* and a thermos full of tea – lukewarm by this time.

Following blue sheep from morning till night, day after day for six months could get a little boring, so I would often mix it up with other data collection, such as sampling for plants or looking for snow leopard signs and scats. The silver lining was the presence of snow leopards. Thanks to the blue sheep's football referee warning whistles, I had spotted snow leopards on multiple occasions. This landscape was packed with blue sheep, and if a snow leopard was out during the day, they would spot it and whistle in alarm. If I worked hard enough, sometimes I, too, would spot the snow leopard. Two other factors in my favour were snow and the slope of the mountains. If there was fresh snow and the snow leopard was on the opposite slope of the valley, then my chances of seeing it would be very high – but only if the blue sheep whistled first. But the most unexpected thing that I ever saw when studying the blue sheep didn't involve a snow leopard at all.

One windy afternoon, when the blue sheep had lain down to chew the cud, I settled between rocks for shelter. Well concealed, as always, for I was hoping to see a snow leopard hunting. I spotted some movement near the cliffs below. The blue sheep were resting about 75 metres down the slope from me, and the cliffs were another 200 metres away. Would this be the day? Unexpectedly, I saw movement at multiple locations. Suddenly, eight Tibetan wolves broke cover, charging up the slope from the cliffs to attack the blue sheep. The largest two led, followed by four other adults – and behind them, at first unseen, two cubs barely able to keep up.

The air was sharp with blue sheep alarm whistles. I was expecting chaos among the herd but, to my amazement, the blue sheep sprang to their feet like cats and dashed directly at the wolves. This made no sense at first, but as the tightly bunched blue sheep made their way towards the cliffs – not allowing the wolves to cut off their escape route – their strategy became clear. The cliffs stretched all the way to the bottom of the Shilla gorge and were a refuge for the blue sheep – who could balance on tiny footholds on their sides – but were more difficult for the wolves to navigate down. The blue sheep ran at an angle to the line of the attacking wolves but towards the same cliffs the wolves came from. It was a race: the blue sheep had to cross the point where their line of escape intersected with the wolves' line of attack. It was like watching a game of American football where two teams charged at each other, the offence trying to get past the defence and the defence trying to tackle the offence. The blue sheep's advantage was that they could go wide on the open expanse of the plateau. The wolves would have hoped for chaos in the herd and that a few individuals would break away and try something different. They must also have known where the blue sheep would run. To my surprise, and perhaps that of the wolves too, the blue sheep herd became a tight bunch and moved at lightning speed. They could not have been more than 10 metres from the leading wolf when they crossed, and the wolves had to chase them back towards the cliffs they came from. Once the blue sheep reached the cliffs and began bounding down the ledges, the wolves gave up. Schaller was right; they were sheep-like, but they were goats. They chose the safety of the cliffs over the opportunity to outrun the wolves on the plateau. In the steep rocky cliffs, the wolves were mere spectators.

By now, the wolves had also spotted me and moved behind me towards the rolling hills, away from the cliffs. Only one wolf stayed back on a crest, his head barely visible, while the others likely rested beyond. A sentry.

I had just witnessed a textbook example of wolf and blue sheep behaviour. It reminded me of famous studies of trophic cascades conducted in the Yellowstone National Park, where in the mid-1990s, wildlife officials reintroduced forty-one wolves after a seventy-year absence. The term trophic cascade was coined in 1979 by the American ecologist Robert Paine to explain how the removal of the starfish (*Pisaster ochraceus*), a predator in the intertidal ecosystem, caused a dramatic shift in the species composition, affecting species beyond its immediate prey, and the impact cascaded throughout the food chain. He called it a trophic cascade when an apex predator, such as wolves or lions or orcas, exerts strong control on the food chain below, affecting the entire ecosystem. Once the wolves re-established themselves in Yellowstone, the returnees affected the population and habitat use of herbivores like elk, which affected how plants and trees regenerated, in turn impacting how beavers built their dams, which led to changes in the course of the rivers and how water was naturally stored and flowed, ultimately benefiting the entire park.

I felt that I had witnessed something similar. These Himalayan wolves were determining where the blue sheep could feed and what part of the habitat they were allowed to use. This could affect the distribution of plant species in the region.

Robert Paine coined another term, keystone species, to describe animals that have a disproportionate impact on their environments and help maintain biodiversity. Removal of these keystone species, Paine argued, would lead to loss of species diversity because only a few competitively superior species would come to dominate the ecosystem. Apex predators functioned as keystone species because they kept the population of dominant herbivores in check, making room for other less competitive species to exist alongside.

In the post-World War II decades, the United States and Europe were marked by rapid industrial and scientific growth. Environmental concerns were non-existent and large carnivores like wolves, bears and mountain lions were at their lowest population levels in both of these places. No wolf had roamed Britain or Ireland in over 250 years; it was extinct in the contiguous forty-eight states of America since 1930, and small, almost ghostly populations survived around the fringes of Europe.

During this time ecologists like Robert Paine were conducting experiments to show that the removal of apex predators and keystone species are damaging to the ecosystem and, more importantly, their restoration can lead to the recovery of these places into healthy and diverse natural habitats. The Yellowstone experiment was monumental in this regard, and it left an indelible mark on the history of large carnivore conservation. Wolves regenerated Yellowstone by changing the populations and behaviour of not only the herbivores but also other co-occurring carnivores. In the few decades since the return of the wolf to Yellowstone, the wolf population has grown to nearly 6,000 across the contiguous forty-eight United States and their populations are considered recovered. In Europe, wolves made a comeback on their own, from little refuges in the mountains in the east in Poland, Romania, Ukraine and in the south in Italy, Spain and Portugal. Every country in mainland Europe reported the presence of wolves in 2022. A similar pattern can be seen with other large carnivores, such as bears and pumas in the United States and bears and lynx in Europe.

When I saw the wolves attacking the blue sheep, I was still young and looking at the world through a lens of awe and inspiration. I only saw what fit the theories I had read in my textbooks. The story of the return of Yellowstone wolves was not just a scientific breakthrough; conservationists were singing paeans to this return-to-the-wild moment. I imagined the wolves and snow leopards playing a similar keystone role in the ecosystem functioning of Spiti.

But even while the wolves had inspired me to think about the trophic cascade and keystone species, I intuited that, relatively speaking, there were fewer wolves in Spiti, and it was the snow leopard that would exert control over this ecosystem. I wanted to study how snow leopards played this function and if they were responsible for maintaining this beautiful ecosystem the way I saw it. I was so much in love with the place, the snow leopard and the idea, that I wanted to prove it. I imagined newspaper headlines declaring: 'Snow leopards are the keystones of the Himalaya'.

The first hurdle in studying this question was to determine the methods to use. We would need to radio-collar snow leopards here in India. Örjan continued to lead the snow leopard collaring program in Mongolia and I continued to learn from his research, and replicating it in the Himalaya would make snow leopards the symbol for this place, as wolves were for Yellowstone.

Putting a radio collar on a snow leopard in India is easier said than done. The same laws that prevent all forms of hunting and trapping of wild animals also prevent research that requires the catching or handling of animals. A radio-collaring research programme would go through the same process of permissions and applications that a proposal for hunting an animal would undergo. Such permissions come from the highest office in the central government, the Ministry of Environment, Forest and Climate Change. I would need permission from the Ministry of Communications to use long-range radio. And I would need permission to use the drugs that Örjan had shown to be safe when handling snow leopards, though it was not yet clear to me which ministry would give such permissions. All these permissions were required at around the same time because they would typically be valid for up to a year, within which we had a four- to five-month window when the temperatures in the Himalaya were suitable to try to catch snow leopards and put on the collars.

A good friend of mine called radio collars the weapon of choice for career suicide. Eminent ecologists and biologists in India had landed themselves in serious trouble (and a few careers were ended) over their research using radio collars. There were three main elements to the story: the first involved a powerful bureaucrat or politician wanting to attend an animal capture and collaring operation along with their entourage. When the researcher denied them that opportunity, keeping in mind the safety of the animal (and the rules of their permissions), the official would take this as a personal insult and the researcher would have made an enemy waiting for an opportunity for revenge. The second element was where the genuine data-driven findings of the research pointed to a need for change in the current management methods. This would irk the local Forest Department manager, and again could create a powerful enemy. The final element involved the death of an animal. The only thing certain after the birth of an individual is its death, but irrespective of how an animal dies, if it has a radio collar or ever had one in its life, powerful enemies will allege that it was the researcher, the radio collar and the capture operation that was responsible for the death of this charismatic and endangered animal. Before the researcher realizes what they are being accused of, they are hounded, their permissions are revoked, and they then must spend years, if not decades, explaining themselves. This was especially true of researchers based at non-governmental institutions who did not have friends within political or bureaucratic circles.

In 2014, I was a prime candidate for this form of career suicide. I was riding a high from having completed my PhD studying the snow leopard. I had accepted a position as a scientist with the Nature Conservation Foundation, the same NGO where I had begun my career as an intern. I also had a dual position with the Snow Leopard Trust (another small conservation NGO), where I supported research programmes in snow leopard range countries, including Mongolia, Kyrgyzstan, China and Pakistan. And I harboured a desire to show that snow leopards were a keystone species in the Himalaya, which needed a radio-collaring project.

Stars seemed to be aligning for the radio-collaring project to become reality. The director of our research programme had painstakingly built good relations with the state Forest Department and the central government. The state government of Himachal Pradesh themselves expressed the desire to start a snow leopard radio-telemetry project. Given the success of Snow Leopard Trust's work in Mongolia and Örjan's expertise, our director led a delegation of government officials to Sweden to meet Örjan and his mentors and colleagues. The visit was a success, and the government allocated the necessary resources for the research project. The next step was for Örjan to lead a team of experts to India, but he and his partner were expecting their first child, so three of his mentors and colleagues would visit us in Spiti to scout for the right locations where we could start the project, with the safety of the snow leopard being our paramount concern.

It was Tanzin Thinley (who had been with me since my first day in Spiti), Ajay Bijoor (the soon-to-be Assistant Director for Conservation), and I who were assigned to accompany the Swedish delegation to the field. These were some of the world's most experienced scientists in leading radio-collaring projects. We were very keen to learn from them. But the alignment of the stars was short-lived, and our troubles started well before we had even come close to a snow leopard.

The Swedish delegation arrived to a quiet welcome in Delhi. We did not want to attract too much attention to ourselves. It would be a two-day drive to Spiti, and the Forest Department officials had provided support along the road, with grand meals wherever we stopped. There was a lot of talk about spicy Indian food and Delhi Belly and how to prevent an upset stomach on a trip to India. We spent two days acclimatizing in Kaza, where, as if on cue, one of the delegation members had an upset stomach. We tried everything to make him feel comfortable, and he tried everything not to make us feel uncomfortable about the situation. The three of them had over a hundred years of fieldwork experience between them, so I was not on my guard, and that was my first mistake. I forgot that they had virtually zero experience working at high elevations. The second mistake was to take the member who was unwell to go scouting in the field. We had carefully identified the sites that we wanted to show them, so that we could choose the best place for the research. I used all my experience from Mongolia to find sites with a similar feel. Places where snow leopards frequently visited and sprayed with their urine to leave pheromone signals. We hardly know anything about this signalling except that in Spiti up to six snow leopards visit such sites in a matter of weeks. Clearly, this is an important form of communication for the members of a species that spend most of their lives alone. The act of spraying a specific spot on a rocky wall requires them to place their feet at precise spots, and that is how we would know where to place our snares to capture and collar them.

The three delegates and the three of us drove out of our base camp after breakfast. It was the sixth day since they had arrived in India and the fourth day since their arrival in Spiti. Typically, this would have been enough time for everybody to acclimatize to the high altitudes of Spiti. But these were not typical times. The delegate with the upset stomach had other symptoms, which he assumed to be due to the upset stomach.

The first spot we visited was only a short drive from the base camp. We had to walk only about two minutes from the road to reach it. The delegates agreed that this was a very fine location to catch a snow leopard for research and they were excited to see more places. The delegate who was unwell was trying to keep up, but he was fading. The second spot needed a ten-minute walk from the road, and we must have walked only five minutes before he started to stumble. We were at about 4,600 metres in elevation and walking downhill. We would have to walk back up, and we thought it best for the unwell delegate to go back to the car with Ajay and Thillay and rest.

The second location was excellent. Our camera traps had found that up to six different snow leopards visited it during one spring and summer season. It was in a place with high blue sheep density. There was a small hut nearby that we could convert into a collaring camp for the researchers and staff to stay in, reducing the time that they would need to reach the snow leopard once it was caught. I could see on their faces that the delegation's work in India was done. They had found the site they were happy with. We walked back to the car, talking excitedly about snow leopards, but then I saw the faces of my two colleagues. The delegate with the upset stomach was not doing well. Ajay whispered in my ears that his speech was slurred. I realized that this was more than an upset stomach, and was instead perhaps altitude sickness. As soon as we reached camp, we put an oxygen saturation meter on his finger to measure his blood's oxygen level. This was well before the Covid pandemic and we knew little about what the different numbers meant, only that 80 per cent was the lowest level that a person might be comfortable with. There was no cell phone reception for 20 kilometres from the base camp. The nearest hospital was 20 kilometres away in Kaza. When the little blue device with its lit green screen beeped, it read 52 per cent. Our base camp at 4,200 metres was too high. The delegate wanted to use the washroom one last time before we headed down in the car and he could barely walk or speak. Ajay stood at the washroom door to make sure everything was okay while two cars waited outside to take us to the hospital in Kaza.

The theory of high-altitude sickness is simple. At higher elevations, atmospheric pressure decreases, so the same volume of air contains less of each of its constituents, including oxygen. At sea level, the partial pressure of oxygen in the air is about 21 kilopascals, and at our base camp, at an elevation of 4,200 metres, it is approximately 13 kilopascals. In common language, it is fair to say that relative to sea level, the available oxygen at our base camp in Spiti is only 60 per cent. But what happens inside our bodies is far more complicated. A healthy person is expected to hold saturated oxygen levels above 90 per cent. If someone's oxygen levels drop below 80 per cent, they need to be treated in critical care.

It took us an hour on the winding single-lane mountain roads to cover the short distance to Kaza. The delegate was admitted to the hospital immediately and I called my sister who was finishing her medical training in Mumbai. When I mentioned that his oxygen levels were 52 per cent, she asked me to listen carefully. She said that at those levels, her hospital would need them to use an invasive ventilator. 'How far is the nearest ICU?' she asked.

My face must have been pale when Ajay looked at me. The doctor said that they had only one compressed oxygen cylinder in this hospital, they didn't have an ICU and, in any case, the patient's condition was too serious to be treated at 3,500 metres in Kaza. She suggested that she could compress one more cylinder and we should drive the patient ten hours to the next hospital in Rampur. This rustic old town was located in a deep gorge, at only 1,200 above sea level.

It was already dusk. The drive she was suggesting would be through the night on some of the most dangerous roads in the world with a critical patient in the back of the car and makeshift breathing apparatus consisting of an oxygen cylinder and masks.

'They don't have a nurse to send with us for the drive. We will be on our own,' Ajay said.

There was another problem. The road to Rampur passes very close to the international border between India and China. A foreign national requires an Inner Line Permit to travel this road, and the official who could provide such a pass was not in town. Several military checkposts line the route and we could lose precious time if we were stuck at one of those. Luckily for us, our director was able to put his government connections to use, and we received a message that all the military posts had been informed of our journey. We considered calling in a helicopter, but that would not arrive until the next morning at the earliest, weather permitting. The doctor said that could be too late. She was hopeful that as we descended in altitude, the patient would already be on the road to recovery.

It was decided that Ajay and Thillay would drive with the patient in one car with an extra driver. I would spend the night in Kaza with the other two delegates and we would leave Kaza the next morning. The two delegates agreed that after seeing the second site, their job was done.

This was a rescue mission. The road from Kaza to Rampur was single-lane and unpaved for the most part. Large sections are marked with signs warning of rock falls and shooting stones. The entire route lies between a deep gorge on one side and a rocky cliff on the other. The roads coil around the mountains like a snake hugging the arm of the charmer. The edges are not marked anywhere along the 250 kilometres and there was no cell phone network for the most part. If the car were to break down, there would be no way of communicating the situation. They would be lucky to average more than 20 kilometres an hour on this dark night.

'Don't worry, the driver knows this road like the back of his hands,' Ajay said as they left.

The other two delegates and I retreated to our rooms. I could not shake the thought that many famous mountaineers and explorers had died of altitude sickness in the Himalaya. Ferdinand Stoliczka was one such early explorer, a Czech naturalist, palaeontologist and geologist in the service of the Geological Survey of India under the British Government from 1862 until his death in 1874. Stoliczka died complaining of headache and trouble breathing in a small village in Ladakh named Murgo, at an altitude of 4,500 metres. He is buried at the Moravian cemetery in Leh. Stoliczka has everything from ammonite fossils to insects, birds, reptiles and even mammals named after him. But none of his experience of explorations could save him from the insidious nature of altitude sickness. It creeps up slowly with mild symptoms of loss of appetite and lack of sound sleep, then develops into headaches. Before long, it generates severe loss of breath and bodily aches as the body shuts down.

As a young student, on one of my early visits to Spiti, I had made the mistake of climbing too high too soon. I was helping another student collect snow leopard faeces samples to study diets. I was on a long ridge and decided to follow it for a little while and see what I found. Our base camp was at 3,500 metres. I did not find much up to an elevation of 4,000 metres, so I decided to walk a little further and, before I knew it, I was at 5,000 metres. The descent was rapid. But I could feel a headache developing. I remembered the lessons from my mountaineering school. My brain was swelling and crushing itself against my skull, and I might be experiencing a cerebral oedema. If I started coughing and spitting blood, that would be pulmonary oedema. I could not eat that night. My head was hurting so much that I pressed it with a hard pillow at first, and then felt a strong desire to put my rucksack over the pillow with added rocks. I felt so miserable that pulling my head out from under the pillow and the rucksack was impossible. I was unsure if I would survive. I imagined that my colleagues would find me dead in the morning. That I could think so clearly meant that I was not as close to danger as I thought. That the delegate had lost all ability to articulate or to even understand where he was tells us how much closer he was to dying.

As the human body moves up in elevation, it needs to adapt in order to draw enough oxygen from the rarefied air to keep organs and muscles functioning. The first thing that our bodies do is to increase the heart rate and breathing to draw more air into the lungs and to pump more blood through them. In this phase, we must keep our demands for oxygen low by not exerting ourselves too much. Over the course of the next few days, our body increases the production of red blood cells to carry oxygen from the lungs to the rest of the body more efficiently. This is the reason athletes train at high elevations – to produce more red blood cells, which will help them later during the actual event. Over time our bodies also increase the number of capillaries to improve the delivery of oxygen to muscles and increase myoglobin – a protein in muscles that stores oxygen – and to make other changes in the mitochondria and haemoglobin to make oxygen delivery more efficient.

This additional production of red blood cells, new proteins and capillaries is energetically expensive, and some of these adaptations are dangerous. A high count of red blood cells and haemoglobin can thicken the blood and together with increased blood pressure, there is an increased risk of strokes and heart attacks. This is one of the reasons that our bodies lose these adaptations when we descend from high places, and we have to repeat the process of acclimatizing all over again when we go back up. But what about people who have lived at high elevations for generations, like the Tibetan people or the people of Spiti? Or, for that matter, animals like the snow leopards?

Tibetan people have lived at high elevations for at least 25,000 years, and modern genomics techniques have found that they have specific genetic adaptations for living in an environment that is low on oxygen. A lower count of haemoglobin keeps their blood thin and flowing well when compared to people from sea level, and they maintain a higher rate of respiration. Two other peoples have lived at high elevations for a long time: the Andean people from South America's Altiplano and the Amhara people from the Simien plateau of Ethiopia. But they occupied these places much later: about 12,000 years ago in the Altiplano and only 5,000 years ago in the Simien plateau. The adaptions of these three different people look somewhat different. The physiology and biochemistry of human adaptation and acclimatization to high altitudes are still poorly understood.

The snow leopard is an even bigger enigma. All cat species have the exact same structure of haemoglobin, and a lower affinity for oxygen, meaning cat species are not as effective at carrying oxygen as many other mammals. What is worse is that the cat's haemoglobin is less sensitive to other factors that help bind more oxygen to it, which is how some species, like humans, control the levels of oxygen being carried to different parts of their body. Cat parents who have moved houses from the seaside to the high mountains have witnessed their pets' struggles trying to acclimatize to the thin air of their new surroundings. So how does the snow leopard cope across a wide elevation gradient from 1,000 to over 6,000 metres? We don't have good answers. We know that snow leopards have mutations in some of the same genes as the Tibetan people (e.g. EGLN1 and EPAS1) but not at the same location within the gene. They seem to have some kind of adaptation that helps them deal with low oxygen levels at high altitudes, but how they do it is still a mystery.

The night I thought I was going to die from altitude sickness, I should have informed my colleagues and let them decide the right thing to do. The complex symptoms of altitude sickness affect our ability to think clearly and make decisions. It is another reason why mountaineers, even experienced ones, make poor judgement calls high up on the mountain when they are struggling with this sickness. Because of this, many expeditions will follow the decisions of the expedition member who remains in the base camp and in control of all their senses. That night, I was concerned that we were too far away from the road for my colleague to organize an evacuation. But most of all, I was embarrassed that despite my training, I had let myself get sick with something that was completely avoidable.

Over the years, my team and I have evacuated many young students, tourists and other visitors who took themselves too high too soon. Often, the people who suffer most from altitude sickness are young men who take pride in their fitness (like me on that night). They believe that their fitness will allow them to push further and harder, and they are embarrassed to acknowledge that they are sick and wait too long to let anybody know. Our sick delegate had done something similar. He was younger than the other two, and he had convinced himself that his illness was related to his upset stomach. By the time he was admitted to an ICU the next morning, he had lost control of his muscles (including bowels, limbs and speech). The doctor was at first not sure if he would suffer organ damage because of acute oxygen deficiency. But he survived, and over several weeks made a full recovery.

This incident was only the beginning of our troubles as far as the snow leopard radio-collaring research programme was concerned. Even when we had permission to catch and collar six snow leopards from the Ministry of Environment, Forest and Climate Change, and permission to use long-range radio (VHF) from the Ministry of Communications, we did not know how to proceed with procuring the right drugs. Our team in Mongolia had tested a new combination of drugs for anaesthetizing the snow leopard, which turned out to be much safer compared with the standard drugs that veterinarians in India were using. However, these new drugs were not available in India, nor were they listed as prohibited drugs. They were simply unacknowledged, which meant nobody knew who was authorized to deal with our request. Around this time, some of the senior officers in the state department changed, and the new guard held our director personally responsible for the opportunities they had missed (such as a field trip to Sweden). They started harassing the director with baseless complaints.

I was swiftly promoted as the Director of the India Programme of the Snow Leopard Trust and my predecessor was asked to take over another project in Kyrgyzstan to remove him from the political trouble. Careers were being damaged before even the first snow leopard was collared and I found myself at the centre of the vortex.

In an ideal world, I should have been happy about the career progression and the opportunity to lead this research programme, but I was filled with foreboding about how badly it could go. If anything went wrong, I would be the scapegoat. The bureaucrats, managers and politicians had immunity for their actions. I, as a scientist based at an NGO, would be the fall guy. Another thing that I had learned from my mountaineering days was that if you keep doing this long enough, something always goes awry. The real question is not whether something will go wrong, but whether you will survive when it does. I switched to survival mode and, after gaining the trust of the team, decided to wind down the radio-telemetry project.

The radio-telemetry project would have answered many fundamental questions, such as the home range size of the snow leopard, the prey density needed for females to have cubs, the rate of movement and the interactions between two snow leopards. Some of these questions could be answered by the research in Mongolia but the deserts of Mongolia were very different from the high-elevation mountain steppe of India. After winding down the project, the only regret I had was that I could not discover whether snow leopards played a keystone role in the high elevations of the Himalaya; whether they controlled the trophic cascades in this place.

Not yet, but we arrived at it the long way around.

The first few years as the India Programme Director were stressful for me. I was now leading a much larger team, the funding requirements were higher than I had raised before, and I was supporting students with their research projects, teaching and managing the expectations of the local government and the central ministry. At this point, I was grateful for the support I received from my team members, the Snow Leopard Trust, my mentors and, most importantly, Ajay Bijoor. He was rock-steady and calm in the face of hardships and maintained a network of hundreds of well-wishers within government offices, community members, fellow scientists and conservationists. Ajay soon took over as the Assistant Director for Conservation in the India team, making it possible for me to focus on academics and research.

It was time to ask the same question using a different method. Instead of collaring snow leopards to study how they affected their prey and how the prey affected the rangelands in India, we decided to study the prey to understand how the snow leopards affected them. In the absence of fine-scale information on individual snow leopards or individual blue sheep, we would have to study this phenomenon at the level of populations.

One of the classical study systems in the field of animal population ecology is the relationship between the Canada lynx (*Lynx canadensis*) and its prey, the snowshoe hare (*Lepus americanus*). These species fascinate biologists because their populations undergo a rhythmic cycle of increase and decrease every ten years, with the hare population reaching as much as ten times its lowest population at the peak of the cycle. The lynx population follows the trend in the hare population with a lag of a couple of years. This is a classic example of dynamic stability – meaning that populations are constantly changing in a fixed and predictable manner without any change in the long-term average. The predation of hares by the lynx is the primary interaction that determines this dynamic stability, and the population of hares determines how much vegetation gets eaten. This is a good example of top-down control of ecosystems where predation by the lynx determines the state of the ecosystem. Experimental removal of the lynx showed that the hare population surged, affecting the entire ecosystem.

Observing blue sheep had often led me to spot snow leopards. It is also how I learnt to read the landscape. The places where blue sheep foraged, their alarm calls, the way they moved in the landscape, the places they chose to bed for the night – everything taught me something about them, their habitats and, crucially, snow leopards. They opened a window into the world of snow leopards.

Ever since we first tried the Double Observer Survey methods to estimate blue sheep population densities, the method has been used widely across mountains around the world to estimate the populations of different species of goats and sheep. It has also been adopted to estimate the populations of critically endangered primates. And every year we surveyed the original five sites that I used for my study on dumplings for snow leopards.

Annually in May, we assembled a team of eight people who went back to the same locations and conducted a Double Observer Survey to estimate again the populations of blue sheep and ibex. Thillay became the leader of this group. Like a sports team that re-grouped every season we would have some new faces (like Munib), while others left, and the rest of us became older and wiser by another year. Each year we added one new data point to trends of blue sheep and ibex populations. Along with the population size each year, we got to know the number of females, kids and males.

Of the five sites, Kibber and Tabo were important because we also monitored the population of livestock in these two locations, and in Kibber we monitored the population of snow leopards. Ever since that first camera trap placed in Kibber gave us the picture of Sunshine and Shadow, we had camera-trapped Kibber almost every year. After completing the Double Observer Surveys for the blue sheep in the month of May, the team would place camera traps for snow leopards. Fifteen days to deploy the camera traps, sixty days of them photographing every animal that moved in front of it and then fifteen days for the team to remove the camera traps. There were years when we failed to do the camera-trapping diligently, either because other areas were to be surveyed, or we were low on camera traps, or natural disasters like a flood kept us from the area. We did not have enough camera traps and money to pay people to undertake long-term snow leopard monitoring in Tabo as well. Sustaining it in Kibber was an achievement in itself. Between 2010 and 2022, we surveyed the blue sheep at the two sites every year and the snow leopard over nine of those thirteen years. Secretly, I harboured the hope that, one day, the blue sheep–snow leopard story would become an ecological classic, like the Canada lynx and snowshoe hare. But long-term data collection is like filling a pot one drop at a time. It cannot be the only thing you do because it is too slow and mundane. Yet you cannot forget about it or the pot will never fill.

During the first few years of the study, when we focused on trends in blue sheep and snow leopard populations in the region, we found that the blue sheep populations in Tabo and Kibber were fluctuating. A pattern emerged where the fluctuations seemed cyclic, similar to the snowshoe hare.

The cyclic patterns of the blue sheep population were clearest in Tabo but I could not explain why and what was causing it, as the snow leopard population remained stable. I became obsessed with this puzzle and posed it to every new member who joined the team. Munib, who worked with me in Kyrgyzstan, was among the first. How can a keystone species like the snow leopard affect blue sheep populations without fluctuations in its own population? Our data from Kibber and Tabo had not yielded beautiful cyclic trends in herbivore and carnivore populations and that frustrated me because it suggested that something other than snow leopards was playing a vital role in the dynamic stability of the ecosystems in the high Himalaya.

Surprisingly, I found the means to the answer in literature. I heard the Nigerian writer, Chimamanda Ngozi Adichie, whose books and short stories I was in love with, speak about the danger of a single story in her TED talk from 2009: 'The single story creates stereotypes, and the problem with stereotypes is not that they are untrue, but that they are incomplete. They make one story become the only story.'

The story of apex predators exerting top-down control, as demonstrated by the wolves in Yellowstone, had become the only story for me. I had failed to acknowledge anything other than the single narrative of the top carnivore as a keystone species. Scientific literature was actually replete with papers on the interactions between top-down and bottom-up forces interacting to determine ecosystem-level outcomes.

Other factors had led to the dominance of the Yellowstone story. A new Western environmentalism movement took shape after Rachel Carson's book *The Silent Spring* was published in 1962. The first-ever United Nations Conference on the Human Environment took place in Stockholm in 1972. Carnivore populations around the world were at their lowest in recent human history and scientists' concern about the extinction of these charismatic beings had become mainstream. Ecologists were looking for arguments to promote not just environmental and species conservation but conservation of carnivores who had been considered enemies of humanity and something to be wiped off the planet. The idea of carnivores as keystone species – as something that plays an important ecological role –immediately grasped the imagination of every person who cared about carnivores.

Once there is plant productivity, whether a place will end up as grassland, savannah or forest is determined by the herbivory it faces, which is determined by the carnivores that affect the population, habitat use and movement of the herbivores. At the scale of hundreds or even thousands of years, the productivity of a place usually changes only marginally, but in modern industrial times the carnivore population density can change dramatically on a much quicker timescale. It took only about 300 years to wipe out the wolf from the contiguous United States after the arrival of Europeans. At local scales, such extirpations can happen within decades, if not years, and have severe consequences for the ecosystem.

However, arguments against these ideas emerged even at the time they were proposed. One simple criticism pointed out that the availability of light and nutrients determine all plant productivity and hence primary plant productivity determines ecosystems. Simple observations bear this out. Highly productive African savannah and Indian forests and grasslands are able to support not only a high density of apex predators but a whole suite of them, creating a hierarchy of apex carnivores. By the same reasoning, extremely low-productivity harsh deserts and high tundra sometimes do not support any species of apex carnivore. This was an argument in support of bottom-up control of ecosystems.

In 2019, four years after I became director of the India programme of the Snow Leopard Trust, the fresh-from-university Dr Manvi Sharma joined our team. Manvi had done her PhD at the Indian Institute of Science. For her PhD she had studied how mosquito larvae responded to predation by dragonfly larvae. Manvi was quiet and soft-spoken. She would keep to herself during general banter but would engage in the academic discussions. Manvi, Munib and I endlessly discussed the trends in the blue sheep populations of Kibber and Tabo until we understood what was happening.

By now, our camera-trap data on snow leopards showed that Kibber and Tabo had similar levels of snow leopard density. The difference between these two places was that Kibber had a large population of livestock, and Tabo barely had any. The mountain steppe ecosystem of Spiti is one of the least productive on the planet. Plant growth is sporadic and slow. The blue sheep in Spiti have to deal with two-fold problems: the harsh winter when there is little vegetation to graze upon and competition for the remaining vegetation with livestock.

When Manvi separated the data for populations of male, female and young blue sheep, she found that fewer young were born in Tabo when the density of blue sheep was very high and more young were born when the overall density was low. From this information we could work out the story backwards.

Blue sheep rut in autumn, right at the beginning of winter, and the young are born in spring when there is plenty of grass for the lactating mother and the young kid. But the mother has to carry the pregnancy through the winter. When blue sheep density was high, there was not enough grass for everyone to eat through the winter and so fewer blue sheep females conceived and carried the pregnancy to term, and fewer young were born the following year. Those few that were born then had to live their first year and survive the first winter in crowded conditions, and so more of them perished during their first winter, fewer blue sheep reached adulthood at year two, and so the total population declined.

The exact reverse happened in years with low blue sheep populations. More females conceived and carried the pregnancy to term, and more of those young survived their first winter and hence greater numbers of young joined the population as adults in year two and this led to a net growth in population.

Together this pattern played itself over and over, creating a dynamically stable blue sheep population in Tabo. It was beautiful.

Kibber told a different story, because of the competition that blue sheep faced with livestock. In Kibber, the density of blue sheep was not the most important limitation on the number of young that were born. Such conditions could be reached any year depending on livestock numbers and where they were grazing. This led to randomness in the data about when blue sheep population in Kibber would increase, decrease or remain stable.

Munib, Manvi and I were mesmerized by how blue sheep populations were responding to the slightest cues of food availability during winter. For the moment we forgot about the snow leopard, a predator that harvested a certain number of blue sheep every year, but at a rate that was merely a background process to that of food availability. For the blue sheep, plant availability created bottom-up forces that affected the recruitment of new blue sheep into the population, and the top-down force of snow leopard predation affected adult survival. In Spiti, the bottom-up forces seemed to trump the top-down forces. In hindsight, this explained my findings in my first snow leopard research project, when I had found that wild-herbivore populations were critical to the size of the snow leopard population.

The humble and almost nondescript blue sheep was in fact the mover and shaker of the mountain steppe ecosystem of the Himalaya – perhaps even a keystone species. Snow leopards ride on the success of the blue sheep in any given stretch of the Himalaya. No blue sheep would mean no snow leopards and more blue sheep would mean more snow leopards. The snow leopard sat at the top of the food pyramid, but their position looked less like a keystone controlling everything below them, and more like a capstone where everything else had to be exactly right for them to be perched on top.

I continue to collect more data, conduct more experiments to test these ideas, and hope that one day the snow leopard–blue sheep system will become a classic ecological pairing, like the lynx and hare.

# 10

# Solar Parks

Only a few weeks after I saw the wolves attack the blue sheep, I was sitting at the base camp alone. This was early in my PhD research, in 2010. The landscape was covered in fresh snow. Most of the team members were busy with their household work. I was staring out of the large glass window, looking blankly at the mountains, when I heard a knock on the front door. Spitians rarely knock, they simply open the door and loudly shout '*Julley*'. It was winter, and we did not expect any non-Spitians to visit Kibber when it was this cold.

Outside was a tall man wearing a heavy black jacket, black balaclava and black glasses, with a thick moustache and a beard. He introduced himself as a photographer who had been sent by the highest offices of the state administration in Shimla to photograph the snow leopard. He produced a letter which said that I was required to help him. I tried to hide my annoyance and said that I could not conjure up snow leopards at will. It might be months, even years, before he saw one. He immediately replied, 'But you have seen and photographed many.' This was only half true. I had seen and photographed snow leopards but weeks if not months passed between each sighting. Clearly these stories were turning into urban legends in Shimla.

We were standing at the doorstep, the vast mountains looming above us. I was unsure if I should invite him inside or find a reason to send him back to Shimla when I looked over his shoulder and could not believe what I was seeing. Over half a kilometre away, what looked like two snow leopards were walking together on the slope of the mountain. He saw the look on my face and followed my gaze but could not see anything. I went into my room and brought out a pair of binoculars, and was able to show him snow leopards, now sitting together on a bed of flat snow overlooking a deep gorge. I went inside to get ready for a trek in the Himalaya on a cold winter's evening. The photographer did not show any surprise when he saw the snow leopards. In his naivete, this was exactly how he had imagined it would all work out.

When I came out of my room a couple of minutes later, I saw him readying a tripod. I advised him that it would be difficult to walk with all that heavy equipment. We would have to plough through at least 2 feet of snow to get close to the animals, and we should be quick because it would start getting dark soon.

I was shocked when he said he was not coming with me. He had a 600-millimetre lens and he was content taking pictures from right where he was. I was stunned and then relieved. I wouldn't have to chaperone him around. I waved a quick goodbye and muttered a couple of sentences about where to sit if he felt like getting warm inside the base camp and quickly disappeared towards the snow leopards to get a closer look. I had not seen one in months.

After half an hour's walk, I sat across a frozen stream from the snow leopards. The wispy clouds scattered the evening light, and the entire Spiti Valley turned monochrome purple. The snow leopards barely took notice of me. They were snuggling next to each other with their long tails in a tangle around their four pairs of legs. Their ochre-grey coats merged into each other and I could not make out where one ended and the other began. They were like two lovers hugging in a single blanket with only their heads sticking out. Surrounded by the beautiful mountains, observing the two snow leopards, I was unable to fathom how fortunate I was to be blessed with a moment like this.

Peak winter is when snow leopards mate. It's the only time of the year when an urge to find other members of their species overtakes these solitary creatures. Snow leopards cannot roar like the other big cats, but their moans travel far. A couple of days ago I had heard the deep and resonating 'auuunh' pulsating through a narrow canyon. Although it can only be described as a moan, the sound was not mournful. Rather it had an urge and a longing in its notes. It repeated a few times with a break of a couple of seconds. Hearing the cry, I had tried hard to spot the snow leopard and failed. After several hours I had left that canyon hoping that his mate had better luck than me in locating him. Clearly, she had found him. They now sat there, 70 metres from me across a deep gorge, nonchalantly throwing caution to the wind in the way only a first love permits.

As it started getting dark, one of the snow leopards peeled away from the other and walked 30 metres into a gully before looking back and waiting. She was now outside the view from the village and my base camp. The light was anyway too poor for the photographer to capture anything from so far away. The other one got up and started walking. He picked up speed as he entered the gully and got closer to his partner and then they both leapt up, meeting each other mid-air and crashing and rolling with the powdery snow lighting up their surroundings in a million glittering crystals. They played with the delight and tentativeness of teenage lovers. Between bouts of tumbling and rolling over each other, they would suddenly stop and stare at each other longingly, then look around shyly or scan their surroundings for intruders before resuming their frolicking.

Unlike other big cats, the difference in the body size between a male and female snow leopard is small. An adult male weighs an average of 42 kilograms, and adult female around 36 kilograms. This small difference was evident in these two.

At a distance, a fox wandered towards them, smelling the ground as he walked. The snow leopards were sitting in a small gully cuddling, one licking the forehead of the other while she closed her eyes in pleasure. I couldn't help but anthropomorphize the fox as wily and cunning, even voyeuristic, although I was the real voyeur. The fox arrived at the edge of the gully, and then he saw them only a few feet away. The yelp that came out of his jaws was the most horrendous and blood-churning noise I have heard in my life, as if the fox had already died and was looking at his own dead body. He gathered himself and ran at a speed that I hadn't imagined foxes capable of. The snow leopards only stared at him in surprise. The male lifted himself on his elbows and raised his brows, and once the fox was out of sight, resumed his attentions to his partner. As the evening progressed, I was straining my eyes to make out the grey snow leopards from the darkness of the impending night. As if I had grown old and my eyesight weak, they faded out of my vision.

I trudged back to camp to find that the photographer had departed, leaving me a note saying that he was extremely delighted with the 'mind-blowing' pictures he got. He must have left when the snow leopards entered the gully and moved out of his view – blissfully unaware of what he had missed.

All this had unfolded near the same place where I had seen Sunshine and Shadow during my first stint here. But unlike that time, I did not go around the village telling everyone about it or showing my pictures. There were no celebrations or parties. I was calm in a way that comes from the wisdom of secret knowledge. I had witnessed something which I believed would become mundane if I tried to narrate it. As if somehow my retelling would diminish the beauty of the memory, as if the images in my mind would be written over with the images of my narration.

This time, I was not cold and frozen as had happened in the past when I watched snow leopards until the last light of the day. I was more experienced. I knew how to stay out of the wind, how to keep my toes and feet moving even when sitting still. The ecologist in me was alert. The fox that jumped out of its skin was a peculiar interaction between an apex carnivore and a meso- (or medium-sized) carnivore. The snow leopard, with its power, was the primary predator. There would be areas in the valley where meso-carnivores and herbivores would be scared to go because they were frequented by snow leopards, something ecologists like to call a landscape of fear.

With the blue sheep data showing that herbivores exerted control of the ecosystem, we wanted to test whether snow leopards exerted pressure on the smaller carnivores. We started working with the records of foxes, weasels and martens that we took as part of our efforts to monitor snow leopards. Every camera trap that captured pictures of snow leopards also took photographs of the other animals that passed in front of it. Together with the PhD students and post-docs in the group, we asked the question: do carnivores avoid each other, as predicted by top-down control? We used sophisticated analytical tools to understand how the different species operated in the space and the time of the day we captured them to reveal how they live alongside other potentially dangerous carnivores. We found that the carnivores did not avoid each other. Some of them actually seemed to be tracking each other closely. The red fox, in particular, appeared to follow the snow leopard. Having considered the results from all possible angles, we discovered that our findings about the relationship between the snow leopard and blue sheep repeated itself for small carnivores.

All carnivores, big and small, seemed to be concentrated in areas they found their prey. Snow leopards around blue sheep, martens around hares and pikas, foxes around hares and voles, weasels around pikas and voles, and bears near juicy, thick, grassy meadows feasting on plant roots and scavenging on the dead herbivores. All the prey species were concentrated in areas of grass. So the smaller carnivore had to take the risk of being around bigger carnivores if it were to eat even though it meant the risk of being eaten. In this landscape, as our earlier conclusions had also shown, grasses and herbivores regulated the ecosystem-level outcomes.

These studies suggest that snow leopards, while being apex predators in this ecosystem, seem to occupy areas with high num-bers of herbivores, who occur in rangelands with healthy plant biomass. The snow leopards do not seem to be affecting where the herbivores graze or the smaller carnivores prey. They are pre-cariously perched as capstones on top of the food chain, and any disturbance to the soil, rain, grasses and herbivores can topple them from this ecosystem. Snow leopard conservation is not as simple as protecting their populations. It means protecting the fragile high mountain ecosystem from the grasses and herbs to the herbivores, so that the snow leopards can survive in this unforgiving landscape.

These findings remain contentious because they are often mis-understood as a claim that snow leopards cannot act as keystone species anywhere. Rather this is only how we understand the interactions between the different species in this less productive ecosystem of the Indian Trans-Himalaya. I am very certain that these interactions would look different in more productive landscapes such as the eastern Himalaya and eastern Tibet, where the snow leopards may indeed play more of a keystone role.

The other reason these findings are contentious is because some critics fear the research will weaken support for snow leopard conservation. I am not concerned about that threat. My view is that because snow leopards are a flagship species which captures the public imagination, in a political as well as a literal sense they remain the top predator. The whole landscape and all the species that live there gain protection from our human interest in the snow leopard.

Top carnivores have been political symbols since the time humans learned to consolidate power. The clash of the British and Russian empires was often talked about as the clash of the lion and the bear. The bald eagle symbolizes the power of the United States. Vladimir Putin presided over a meeting of tiger biologists and conservationists in St Petersburg in 2010. India's prime minister, Narendra Modi, did the same in Delhi in 2016. Almazbek Atambayev, then-president of Kyrgyzstan, one of the world's most mountainous countries, led one of the largest meetings of snow leopard scientists, conservationists and policy makers from around the world in 2017. Atambayav hoped that both he and his country would be associated with the majesty and power of the flagship species of the high mountains of Asia.

One morning in 2022, Ajay shared a newspaper article which claimed that one of India's largest solar power plants was to be built in Spiti. The plant would generate 800 megawatts and be comprised of thirteen different units of solar panels spread over large tracts across the region. Some of the names of the pastures that would be covered under the panels were all too familiar for me and the team. They have appeared repeatedly on our data sheets over the years. Camera trap from Dombachen, snow leopard scat from

Paldar, blue sheep herd of Lungwooh, plant samples from Pangmo. And now, the newspaper article said, there would be a solar park in Dombachen, Paldar, Lungwooh, Pangmo and nine other places.

Dombachen was where I had conducted my master's thesis and followed the blue sheep for a whole winter, studying their every move. Every plant they fed on, every tactic that the males employed to woo the females during the rut, every nook and cranny that the females used when they gave birth to the new kid in spring, and every kill the snow leopard made. Here I had seen the wolves attack the blue sheep, and the blue sheep score a touchdown against the wolves. Here I had seen Shadow materialize from behind his mother Sunshine. And I had seen Sunshine raise another litter of two cubs after Shadow dispersed to find a territory of his own.

Now I imagined a dystopic future with black silica panels and grey steel poles repeating endlessly over the horizon. The bare ground underneath where the sun would never reach. Rats and pikas scurrying about, feeding on the weeds that would grow around the edges. Empty skies because the golden eagles would not be able to hunt the rats through the panels. The glass of the panels blinding the eagles as they swooped down. The blue sheep forgotten as a distant memory of the past and the snow leopard a literal ghost of the mountains.

Solar power is green energy. Climate change is the larger beast threatening the existence of not just the blue sheep and the snow leopard but many more species including humans. Some of the villages in Spiti had already consented to solar parks in their range-lands in the hope of better employment opportunities. At the same time, critical assessment of large solar power parks from other places in India had shown that they were not as efficient in producing energy as originally claimed, nor did they generate as many jobs. Once construction was completed, only a handful of people oper-ated solar plants covering several hundred hectares.

Hydroelectric power, another source of energy that does not emit large amounts of carbon into the atmosphere, has been controversial in India, especially in the Himalaya. These large projects have typ-ically benefited only the project proponents at the cost of traditional local livelihoods. The Himalaya are seismically active and the threat of a large earthquake looms. If it hit one of the big dams it could take a large number of human lives. Glacial lake outbursts are one of the more severe kinds of environmental disasters in the Himalaya, and these floods gather more water as they go and breach hydroelectric power plants along the river. In peninsular India, dams like the Sardar Sarovar on the Narmada River submerged hundreds of square kilometres of forests and thousands of forest dwellers' lives. These people were forcefully moved from their homes of several generations.

Kinnaur district on the east of Spiti has one of the highest concentrations of hydroelectric power plants anywhere in the Himalaya. These dams have not only destroyed all the aquatic life in the river, but killed the River Satluj. Now, the water flows through a series of tunnels from one power station to the next. A trickle is allowed in the river to preserve the minimum ecological flow. The Kinnaura youth have started a movement against the large hydroelectric project in their region. One young person shouted, 'We don't want dams, why don't you understand, no means no,' and the phrase 'No means no' caught on as the protest slogan. The dams led to an increase in landslides because construction had loosened the mountainsides. In 2021, two landslides claimed twenty-two local lives in an area where landslides were rare. The local people pointed to the dams. They also pointed out that local freshwater springs had dried up after tunnel construction moved the river water to the power generation centres. While the power generated here is supplied to distant cities like Delhi, it has done little to improve the local economy and ecology.

People in Spiti had been saved from these pressures because they were higher up in elevation and these large hydroelectric projects would never reach them. Now an enormous solar park was proposed in their backyards, and it was unclear how it would affect the local economy and life in the region.

The climate movement globally often presents as an intergenerational conflict, where the younger generation demands action from global leaders born at an earlier time and who view the impacts of climate change differently. People my age find themselves somewhere in between, but now I was faced with a difficult choice. My team and I turned to data to help us understand the impact and how we would respond to the news. We made a distribution map of the best habitats for snow leopards in Spiti and plotted the proposed solar power plants on top of it. We found that at least half of the panels would sit on some of the best snow leopard habitats in the country.

What was our moral responsibility? It made sense to us that solar power generation needed to be decentralized and made into rooftop projects. These large plants took up too much land and also used large volumes of water to clean the panels. But how would the younger generations judge us for not seizing the opportunity to generate green power? How would my daughter see me in ten years' time? Would she see me as someone who tried to save the snow leopard or as someone who was opposed to ways of reducing carbon by moving to greener sources of energy?

We decided to publish the information we had on the overlap between snow leopard habitats and proposed solar power plants in Spiti to generate debate. We identified the locations where solar parks could be built without serious damage to snow leopard habitats while pointing out those that would seriously affect the snow leopard population.

By now, we were certain that the bottom-up forces were stronger in Spiti than the top-down effect of the predation by snow leopards. We knew that even if solar parks were set up in areas less suitable for the snow leopards, the presence of the solar panels would push all the livestock into the few remaining pastures. These would be decimated by overgrazing, which would affect blue sheep populations, and no blue sheep meant no snow leopards.

The trade-off between snow leopard conservation and solar parks is only the surface of the problem. Each of our conservation actions, like building livestock corrals, supporting community livelihoods and taking a stand against mining and solar parks, is geared towards addressing a specific problem at a specific place, while climate change is insidiously wrecking the entire planet.

A year later, a PhD student came to my office and told me about a problem in the Great Himalayan National Park. Many of the sites where our camera traps had photographed snow leopards in 2019 were now full of images of the common leopard. Unlike Ladakh, which lies in the rain-shadow region in the north of the Himalaya, the Great Himalayan National Park is on the southern slopes of the Himalaya. In one steep climb, this park goes from 2,000 to 6,000 metres in elevation. The common leopard with its sunshine-yellow fur and contrasting black spots arranged in flower-shaped rosettes is more than twice the size of its smaller cousin, the snow leopard. Common leopards usually keep to the forested habitats but with the rising temperatures of each passing summer they are moving higher in the landscape. This is dangerous for the snow leopard population since common leopards threaten them as the apex predator in the areas they range. We sat in silence for some time as we contemplated the problem. We both knew what we would be researching in the near future.

Interaction with the common leopard is just one of the many pathways by which climate change will impact the snow leopards. The high mountains of Asia are often called the 'third pole' of the world because of their large mass of ice and glaciers. Warming and melting in this region will have global impacts, but how exactly they will affect the snow leopard can only be guessed. In 2015, we set up fibreglass chambers at rangelands as high as 4,800 metres in Spiti and Mongolia to simulate a 1° C rise in temperature and, separately, 50 per cent drought conditions to test the independent effects of both these treatments on the vegetation of the region. After three years of monitoring these experiments we found that even a small amount of warming will reduce the vegetation cover of the region by a third and persistent drought could reduce it by half, across both the countries, which could affect the wild herbivores and have implications for snow leopards.

Our research showed us that snow leopard conservation needed landscape-level protections. A large solar park would severely squeeze the blue sheep and snow leopards for food and space and exacerbate the effects of climate change in the region.

While we thought through the ecological, climate and moral arguments, Ajay shared another news item which announced that the Spiti solar plant project had been shelved due to its lack of feasibility. We were greatly relieved, but also knew that a project of this proportion would not stay shelved forever. It will resurface in a few years with new calculations for feasibility. We had to use the time to put together the best possible science to make it possible for snow leopards to survive in a rapidly changing mountainscape.

# PART IV

# Answers

# II

# Have You Seen the Snow Leopard?

February 2023 was an unusually warm month in Spiti. There was barely any snow on the ground. I sat in my room in our base camp in Kibber village, attending a video call on Zoom. Internet and cell phones had made their way to the village a couple of years ago. Electricity too was more reliable, and I now used an electric heater rather than the wood-burning stove that had kept me warm for over a decade. The locally made red carpet shone brighter without the wood smoke.

Thinley-ji walked into the room with two cups of chai and nonchalantly said, 'A sighting is happening. It's the Mother and the Siblings.'

'Same place as yesterday?' I asked, keeping my eyes on the screen.

'Yes, a couple of hundred metres east of where they were seen yesterday evening. Exactly where we saw Sunshine and her cub in 2008.' That comment brought a flood of memories. Back then, it was completely unexpected to see snow leopards out during a bright sunny day and so the first person to notice them had thought it was a blue sheep struggling in deep snow. After spotting them from the terrace, my colleagues, friends and I had rushed out of the door, barely taking our jackets. The sighting of Sunshine and Shadow had been only the second sighting of a snow leopard in a decade for the researchers then present in Kibber. The whole village had partied all night. Today, two sightings in two days of winter was considered par for the course.

Our long-term camera-trap data showed that the population of snow leopards had remained pretty much the same over the past decade and a half, but through continuous research and monitoring, we understand the snow leopard better today. We know that their camouflage fails against the winter snow and it becomes easier to spot their silhouette against bright snow. It is like taking away their invisibility cloak. In winter, their pug marks can be followed with a pair of binoculars for miles around. Over the years, the quality of and access to binoculars, spotting scopes and long-lens cameras has improved. The road leading up to Spiti does not shut down for winter any more, and the influx of tourists in winter has created a new economy for snow leopard-spotting. Now, ten young men from the village spread out to look for snow leopards every morning. With the first sign of a snow leopard, radio messages start cracking across the village. The tourists then trudge from their warm cottages to get a peek at the snow leopard through a spotting scope already trained on the creature. Spotters are mindful to keep enough distance from the snow leopard and usually stay across a gorge so that the snow leopards don't feel threatened by the people. Seeing the snow leopard from across a gorge also makes for a great viewing experience because the snow leopard is often at eye-level and the tourist doesn't have to crane their neck to see them.

A similar increase in sightings of snow leopards has been found in other places such as the valley of cats in China and parts of Ladakh. Many more places would like to join this list but regular sightings are only possible through a coming together of a sizeable population of snow leopards, knowledge of their habits, access to the place in winter, and a committed group of local people to track them.

It was my last day on this field trip in Kibber before I headed back to Bangalore. I decided to observe the snow leopards one last time, after I finished my video call. Thinley and I reached the place where we had watched Sunshine and Shadow fifteen winters ago. This time, there was little snow. Instead, we were confronted by a line of spotting scopes and long-lens cameras perched along the brow of the cliff, just above the steep 110-metre drop to the Samba-Lungba stream. The snow leopards were across the gorge a few ledges below us. The mother, who was simply named Mother, was curled into a ball of fur looking exactly like a round boulder. The Siblings, as her two cubs had come to be known, were below an overhanging rock and difficult to see from our position. The tourists had been there for over three hours, waiting for the snow leopards to move, but with the animals asleep, none were at their scopes or cameras. Behind was a row of picnic tables equipped with thermoses of tea and coffee, surrounded by hundreds of people holding steaming cups and chattering about their latest wildlife adventures in Africa, Antarctica or the tiger heartland of Central India. The place resembled a country fair. New arrivals would never have guessed that three snow leopards slept only 200 metres away. When the snow leopards woke up, they would be able to see the line of tourists on the cliff but not the chaos behind them. But even if they could, our past experience had told us that they would not react as long as the tourists stayed on their side of the gorge. Over the last decade the snow leopards had become very tolerant of people in this part of Spiti where they had not experienced any harm from being watched.

Amidst the throng, I spotted Prasenjeet Yadav, my old friend who I'd worked with in Kyrgyzstan. He was making a snow leopard film for the BBC. Three different companies, he told me, were filming the same snow leopard today. Thinley, Prasen and I were deep in conversation when someone suddenly said, 'She is moving,' and next came a scattering of feet on the hard frozen ground and the shuttering of cameras. Right now, people-watching was more fun than watching the snow leopards. But the camera shutters soon fell silent, as the Mother lifted her head, looked around and went back to sleep, and people returned to their picnic tables.

Solitary, ghostly, shy, alone, hard to see and even lonely; these are some of the words that are often used to describe the snow leopard. But the Siblings were none of those things. They were oversized kittens, the size of small golden retrievers, and bold, perhaps never having experienced trouble from people. They acted goofily, and played with everything from sticks to dried cow turd. Their movements lacked the finesse of their mother's. They were clumsy and uncoordinated like novice ballerinas.

At around 4 p.m., the Mother and the Siblings twitched their ears, and the tourists and film-makers clicked away. This ear twitch was the most exciting thing that had happened all day. Then they woke up and stretched like house cats, and excitement built in their audience. A cub walked a few steps and hunched down to poop – only cats can look cute while pooping. I made a mental note to go down to the spot later and get a sample for genetic analysis.

A half-frozen stream flowed past the three snow leopards. Across a bend in the river, on a small grassy meadow, a herd of ibex were grazing in the golden light of the evening sun. We could see the ibex and the snow leopards, but the ibex could not see the snow leopards.

The Mother, however, seemed to know about the ibex, and she started walking deliberately and carefully towards them. She could not have seen them because of the bend in the cliff along the river so she either heard them or knew that they grazed there. The Siblings followed her with equal care and attention. I was impressed with the discipline the cubs were exhibiting. As they came closer to the ibex, the Mother shifted into a stalking mode, and the cubs followed. This was the most important thing they would learn if they were to survive in the wild.

Tension built up in the snow leopards, and we witnessed it from across the gorge from an overlooking cliff, as if watching a nail-biting cricket match from the high stands. Now, the Mother was within thirty metres of three large male ibex. With their scimitar-shaped horns these goats are formidable prey even for a veteran snow leopard mother. The Siblings stayed back and observed. I wondered how she communicated to them that they needed to remain behind her.

She got closer, 20 metres; the cubs were watching, and the viewers on the bank were holding their breath. Prasen had his BBC camera on the tripod and he worked it with the precision of a sniper. The Mother could attack at any moment now. It would be as spectacular as a cheetah coursing down an impala in the Serengeti. Now, 5 metres, and the ibex had still not seen her, and she was only one leap away from making a kill. She had approached the ibex head-on and now she would have to be careful to avoid the big horns which can cause a serious injury. She would make her move once she had a better angle to lunge at their neck or shoulder.

At the moment we expected her to commit to the leap, the Siblings became jittery and started chasing the ibex herd in the open meadow 40 metres away from them. The element of surprise was lost. The tension disintegrated into chaos; twenty-five ibex were running around and two one-year-old snow leopards bounded clumsily after them. The Mother sat glumly. The cubs had spoilt her hard work with their impatience.

After the dust had settled, the Siblings approached their mother with tentative steps. I was expecting her to reprimand them, but she was their mother after all. She licked them clean and they walked away into the dusk that was spreading over the mountains.

The Siblings went on to attain celebrity status, their exploits tracked on social media through the tourists recording their movements. In another year they separated from their mother but stayed together until their third summer. Their boldness and lack of fear of humans gave us great insights into their methods of hunting. They would corner ibex and blue sheep along the edge of a cliff and in desperation the herbivores would fall to their death. The snow leopards would then walk down the long and easy way knowing that a meal would be waiting for them at the bottom. Watching an ibex fall off a cliff as tall as a 40-storey building was gut-wrenching. Tourists held their breath, fearing that the ibex would drag one of the Siblings with them, but the snow leopards used their leathery, wide paws to gain a firm grip. Their strong chest and thigh muscles worked like anchors. When the third summer approached, the Siblings disappeared; dispersed to find new homes, just like Shadow did a decade ago, and M9 in Mongolia a few years later. New snow leopards from neighbouring territories moved in and new observations are being made by the scientists, local people and tourists.

'Have you seen the snow leopard?' Peter Matthiessen was asked

by a journalist. 'No!' he replied. 'Isn't that wonderful?' This response is as evocative today as it was fifty years ago. Matthiessen's
book *The Snow Leopard*, about his expedition with Dr George
Schaller to the Shey Monastery in Nepal, merged the adventure of a
Himalayan expedition in search of the snow leopard with a spiritual
search for the self and turned the trek into a pilgrimage. While
Matthiessen missed seeing the snow leopard, he seemed to have
found a meaning to being through the journey. Since then, seeing a
snow leopard has meant many different things to different people,
just as climbing Everest, circumnavigating the globe, or walking the
El Camino have helped people make sense of their world.

Back in the 1970s, snow leopards lived in a remote and isolated
mountain kingdom, and tracking them required the seeker to undertake a physically demanding and arduous journey for the outside
chance of glimpsing the world's most elusive cat. Since then access
into this area has become much easier, and all-weather roads and
airports deep in the mountains have reduced the hiking time to the
best sites to find snow leopards from many weeks to almost nothing.

In 2014, I received an email from a young undergrad student,
Gopal Khanal, from Nepal. He wanted to work with me on snow
leopards eventually, but at the moment he had data on the Ganges
river dolphins (*Platanista gangetica*), of which only a handful were
left in Nepal. Although I had never studied a marine mammal
before, I had used the same analytical technique that Gopal wanted
to use. This was the beginning of a chain of events that brought
me close to the route and experience of Peter Matthiessen's quest
for the snow leopard.

Gopal was a short, stout and strong man with broad shoulders.
He was always clean-shaven but his stubble suggested that he could
grow a thick beard if he chose to. He was always smartly dressed
and he was very well-read. When he met me the first time, he had
already read all the research papers that I had written. He wrote up
a research paper on the river dolphins with me and was ready to ven-
ture into the mountains in search of the snow leopard. I suggested
the master's programme in Wildlife Biology and Conservation at
the National Centre for Biological Sciences, my alma mater. This

was where my dreams had come true and now Gopal would have the same opportunity.

Gopal excelled through the first two semesters of the course. He wanted to do his thesis on the snow leopard in no other place but Upper Dolpa, the area surrounding the Shey Gompa. In June 2017, to lay the groundwork for his upcoming fieldwork, Gopal invited me to walk with him in the steps of Peter Matthiessen and George Schaller.

'Will we get to see a snow leopard?' I asked and he only smiled in return.

Tucked in the north and west of mighty Dhaulagiri (8,167 m), far away from the hustle and bustle of Kathmandu, Pokhara, Everest and Chitwan National Park, is the high-elevation plateau of Dolpa where Shey monastery (or Crystal monastery as Peter Matthiessen and George Schaller called it) is located.

The Dolpa region has two distinct halves. Lower Dolpa is on the southern slopes of the greater Himalayan range and is covered in conifer and birch forest to an elevation of nearly 4,000 metres, after which are found alpine meadows, snow, rock and ice going well over 6,000 metres. The picturesque Phocksundo Lake is situated at an elevation of 3,600 metres. Upper Dolpa lies north of the greater Himalaya beyond the Kang La Pass. This region is more of an extension of the Tibetan plateau. Being in the rain shadow of the Himalaya, it is covered in grasses, low shrubs and bushes. The Shey monastery overlooks wide rolling hills covered in emerald-green grass in spring and golden honeydew in autumn before being completely covered in white, powdery snow in winter. The entire region was declared the Shey-Phocksundo National Park in 1984; at 3,500 square kilometres, this is the largest national park in Nepal.

George Schaller was attracted to Dolpa because of the ease of observing blue sheep here and the relative habituation of these animals to humans visiting the monastery. Gopal wanted to be among the first Nepali scientists to study the snow leopard and blue sheep in the land where Schaller had studied them and where Peter Matthiessen wrote the legend of *The Snow Leopard.*

Working with Gopal on his master's thesis brought me full circle. I felt fortunate to be able to help him the way Charu had helped me over ten years before. The Dolpa region of Nepal was similar to Spiti geographically and ecologically and we could translate many of our hard-learnt lessons to his fieldwork.

After two days of driving from Spiti, then one night on the bus from Delhi, I crossed the international border between India and Nepal at Nepalgunj. Another overnight bus took us to Jajarkot, and finally it was time to start walking – or so I thought.

Usually, tourists can take a flight directly to the Dolpa airport in Donai (the gateway to the Shey-Phocksundo National Park), but the airport was under renovation, and all flights had been suspended for a few months. I was secretly happy, both because the aviation industry in Nepal had a very poor record and ranked among the most unsafe places in the world to board an aircraft and because I wanted to walk the mountains. I believe that walking is the only way to understand a mountain. I wanted to experience what Schaller and Matthiessen had experienced.

From Jajarkot, we crossed a bridge under construction over the Bheri River and walked along the road which would connect the national park with the rest of the country. None of the bridges were ready yet so the road was mostly free of vehicles.

A few enterprising young men had managed to get run-down old vans across the rivers onto some of the long continuous stretches of the roads and plied them as transport services. These men instantly rose in importance in the local economy and started controlling key businesses, legal and illegal. The local system was so utterly dependent on them that the drivers assumed mafia-like arrogance. We were waiting for one of these vehicles to take us across a long stretch of 20–25 kilometres of road to the next river and I asked the driver who was sitting in a shack nearby how long it would be before we departed.

'When I am done drinking all the beer I want,' he responded in broken English, recognizing that I was a foreigner. I could see Gopal's face fall into embarrassment. In my head, I was trying to calculate the probability of dying in an air crash versus a car crash.

It was close to the evening when the vehicle started to move. Two young men, barely out of their teens, were in command. Both were drunk like sailors on a holiday, but their driving was surprisingly smooth. That I was a father of a 2-year-old daughter played on my mind. And as the trip progressed over the next few days, my fear of dying on the road completely took over my experience. More so than the snow leopards, Gopal's upcoming research and walking the footsteps of Schaller and Matthiessen, I became transfixed by fear of an accident. We were barely doing any walking anyway.

We had only been travelling for an hour, and it was already getting dark. Gopal had told me we had a couple of hours ahead of us when the two men pulled up next to two shacks. It was dinner time, they announced. Out came a few more beer bottles and the duo with a couple of other passengers started drinking again. Gopal ordered some food for us while I tried to stay calm. A middle-aged couple cooked at a wood-fired mud stove. In the traditional manner, the woman had a smear of vermillion in the parting of her hair. The man wore at least two layers of trousers and jackets, the top layer caked in kitchen grime and dust. A young girl, barely in her teens, did all the chores. She never stopped smiling and seemed lost in her thoughts, humming to herself even as she was ordered around. There was no sense of annoyance in her demeanour. Other passengers sat bored. I tried to stay busy by making a note of the vegetation and the birds that I could identify. Two barn swallows flew in and out of the shacks.

I noticed an infant sleeping on a mat next to the kitchen. He had rolled off the mat into the mud. He must have been only a few months younger than my daughter and memories of her flooded my mind. I wanted to pick him up in my arms and show him the swallows. When he broke out in a feeble cry, I expected the woman from the kitchen to take the child but I was shocked to see the young girl move the infant to her breast nonchalantly, as if waiting for this opportunity to drop everything else. Her child. She was a child herself, and could surely not be a mother at the same time. She held the child like a little girl in a frock cuddling her pet kitten.

I was experiencing a strange rage and pity at the same time. This was not the first time I had been confronted by such poverty; I had seen it in my own village, but distance from my own daughter, the arrogance of the two drivers and the helplessness of the other passengers stirred up a cocktail of emotions that made me nauseous. Gopal sensed my unease. He told me how many young men from the region had gone away to work as contract labourers to build stadiums for the upcoming Fifa World Cup in Qatar. Their young brides and kids were left behind, sometimes with grandparents and at other times to fend for themselves. Some of the men sent money home, and some would die working long hours in extreme heat. The young woman running the neighbouring shack had not heard from her husband in many months. Nobody knew if he was alive or dead. Two kids, barely 5 or 6 years old, darted in and out of her kitchen, sometimes hiding behind her long round skirt. The two drunken drivers walked over to her shack and started soliciting her, right there in front of her children.

I was burning in anger but I did not want to get into a fight. The woman had been here for a long time and seemed capable of handling the situation, or so I convinced myself.

I told Gopal that I could not get back in the car with these two maniacs, without one of us killing the other. We had sleeping bags and could find a spot and stretch out somewhere on the hillside. By now, our drivers had created a scene. They were shaming the young woman for thwarting their advances. Gopal checked with the family who had cooked our dinner, and they offered us a room for the night, but they did not want to antagonize the drivers: 'They are our regulars.' The drivers threw a fit when they heard that we were not getting back in the car. Gopal paid and said we were sleepy and did not want to continue, but the drivers were not buying our excuses. They left shouting curses.

The peace and tranquillity of the mountains returned. For the first time, I noticed the bright moonlight and the broad-leafed oak silhouetted against the sky. The old man led us on a small single path through the woods and we came upon a little barn-like room which smelled of stored grain and jute sacks. The teenage mother was making two beds by the light of a candle when our torches lit up the room. She was still humming to herself and smiling.

Sleep eluded me for a long time. Gopal's voice came from the darkness. 'I think the drivers will talk about us and nobody will drive us now. We may have to walk the rest of the way and it will add two days to our journey each way.' I was genuinely happy that we were finally walking. It would cost me precious time, but at the moment, I was relieved about not having to get into a tin box with wheels driven by men I didn't trust. I lay there awake, thinking about what Peter Matthiessen and George Schaller would have made of the situation.

The next day we ambled along the road towards Dunai. We brought out our binoculars and saw a spot-winged myna, a new bird that I had never seen before – a lifer. The mountain slopes were denuded, with only a few large tree trunks. The rest was covered in weeds and grasses, fresh from the recent rains.

Soon, the road filled with people walking towards us. People carrying pots and pans and sleeping mats. People with tired faces and blank stares. Their walk was purposeful and yet their feet trudged along. The children walked ahead, often dressed in fashionable clothes and colourful shoes, and the parents and the grandparents walked behind with large backpacks and head-loads. Many of them had camped along the roadside and were packing up and starting their day.

'This is the end of the yarsagumba season. They are going home from the Shey-Phocksundo National Park after collecting and selling the yarsagumba,' Gopal told me.

Yarsagumba in Tibetan means 'summer grass winter worm'. It is a fungus, *Ophiocordyceps sinensis*, which infects the caterpillars of the ghost moth. It looks like a yellow-green or brown-black stalk growing out of the head of the caterpillar. The stalk grows out of the ground like the first shoot of a seed, while the dead caterpillar stays vertical in the ground like the roots of a new plant. The mystical half-animal half-plant appearance of this parasitic relationship between a fungus and an insect has led to many myths in Tibetan and Chinese medicinal texts and, like so many other animal products, it is valued as an aphrodisiac in Tibetan and Chinese medicine. Although the yarsagumba is mentioned in historical Tibetan texts as far back as the 1400s, a surge in demand occurred after 2000. By 2012, the best quality yarsagumba fetched about $50,000 per kilo, making it one of the most expensive animal products in the world.

This caterpillar fungus is only found in the alpine meadows above an elevation of 3,500 metres on the Tibetan plateau and the Himalaya, mapping almost exactly onto the distribution range of the snow leopard. With rising demand and prices, hundreds of thousands of people from the lower Himalaya make the trek every year to the alpine meadows in the months of June and July to harvest the caterpillar fungus, and hundreds of tarpaulin tents pop up in prime snow leopard and blue sheep habitat.

As we trekked up, hordes of people walked down and we noticed makeshift enterprising businesses that had popped up on the road near small mountain villages. Small hotels served meals and alcohol, and gambling houses promised entertainment and excitement. Gopal was reluctant to discuss these unpleasant spectres in his beautiful country, and I did not want to put him in an awkward position by asking. But I observed what I could, and Gopal and other people offered some information voluntarily.

'Yarsa collection has wrecked the national park,' Gopal explained. Every summer, about ten thousand people spread over the entire park and scour the place inch by inch in search of the caterpillar fungus. To find the fungus, a worker must lie on their belly and poke the grass blade by blade. When they locate some fungus they must gently pluck it out with a knife with the caterpillar still attached to the stalk. 'All these people need food, and they put out snares which continue to kill animals like the blue sheep well after the people have left,' Gopal said.

'This is like the Gold Rush,' I offered.

'Yes, the merchants fly into the park in their helicopters, pay the people in cash, and fly out with the yarsa. The people walking back carry cash home and the gambling houses try to lighten them as much as they can. This is also why the women and children come along with the men, so they can ensure that the money reaches their homes. Last year, a fight broke out between two groups inside the park over access to different meadows and nine people were hacked to death with kukri knives in a single evening.' Gopal spelt it all out. There was a sadness in his voice that comes with the relief of having spoken a difficult secret about oneself.

'The park is still very beautiful but there is an ugliness surrounding it,' he said with melancholy.

Gopal pulled his phone out of his pocket and showed me a grainy video. In it, a large group of men had gathered in a verdant green meadow. When the frame stabilized, I could see a man pulling the tail of a snow leopard who was trying to get away from the crowd, struggling to find its feet. A normal healthy snow leopard would take a swipe at anyone tugging its tail but this one was struggling to hold its balance.

'This is what they do for entertainment. It's mob psychology. They probably laced a bait with some drug and then tortured the snow leopard for fun,' Gopal said.

People took turns harassing the poor animal and, after a point, the snow leopard gave up its efforts to get away and resigned itself to its fate. The video ended abruptly, leaving me to imagine the worst. I felt as nauseous as if I had seen gory footage of a dead body. My body was reacting viscerally and I did not know how to control it.

Soon the broad jeep track gave way to single trails and the methods of constructing the road became more visible. The trees were being cleared and heavy equipment blasted through the rocks to widen the trail. Some of this blasted rock was pushed down towards the stream while rocks that could be used as construction material were ferried by mule trains. Every so often a long line of a hundred or two hundred mules walked past us in the opposite direction. We were careful to always step onto the mountain side of the trail to let them pass. Moving to the cliff side of the trail meant that even a gentle nudge from one of the unruly mules could plunge us into the abyss below. Every so often we came to a bloodied spot on the trail where a mule had broken a leg in a rock crevice and been stampeded upon by the other mules in the train. We ended the difficult day in a small village where Gopal found us a room in the house of a kind family with a small orchard in the garden.

Suli Gad was the first place our path met Schaller and Peter Matthiessen's. They had arrived here from the south-east, crossing over a high pass near the 8,000-metre-tall Dhaulagiri. At Suli Gad, we sat in the shadow of Dhaulagiri but I had not yet had a clear view of the mountain. We were too close to the mountain to be able to see the magnificent cone of its icy peak; we were almost in the belly of the hills surrounding it.

The next morning, we continued our journey along the stream that directly flowed from Phocksundo Lake. We would trek along it for two days to reach the turquoise blue waters. The hiking path wandered through a deep-green forest of walnut and cedar trees. Stretches of the ground were wet and cold where the sun had not reached the floor all summer. The trees muffled the sound of the raging stream which was only a five-minute walk away. This was where Matthiessen had seen an Asiatic black bear and my eyes darted from rock to tree stumps hoping to make out the shape of a bear. We had been teleported into a different world after we crossed the checkpost at Suli Gad. For the first time on this journey, Gopal and I were alone. No mule trains, nobody blasting the rocks to build a road, and the yarsa collectors had moved past us. Only graffiti on some of the large boulders reminded us of the civil war that had been fought here a decade ago. For the first time in days, I was walking with my own thoughts.

'I wonder if anywhere on earth there is a river more beautiful than the upper Suli Gad in early autumn,' Matthiessen had written about this place forty-five years ago. It was not quite autumn yet, but I agreed with him. I felt his presence as we walked on. I would have liked to have known what he thought about our journey leading up to Suli Gad, about the yarsa collectors and the women running the roadside shacks.

I never got to meet Peter Matthiessen. I would have very much liked to. But I was fortunate to meet George Schaller. The first time I met him I was a master's student and Schaller visited our institute to give a talk. I remember embarrassing myself during the public question and answer session by asking him if he found the Virunga mountains more enchanting or the mountain gorillas. Only after the words had escaped my mouth did I realise the absurdity of the question. I don't remember his reply, just the cold stare of the head of the institution looking back at me from the front row.

My second meeting with Schaller was on the Qinghai–Tibet plateau in 2019. The Chinese government had organized a conference in the city of Xining, the capital of Qinghai province, to discuss plans to create a US-like network of national parks. Many senior party officials were in attendance. The rumours were that the provincial governor might be present. After the publication of Xi Jinping's thoughts on China as an ecological civilization there was a push for greater environmental conservation in China. Only a handful of foreigners were invited, including a former director of the US National Park Service, George Schaller and a few others who worked at Chinese universities. I was not an invitee but I was in the country for another meeting relating to the conference. A small gathering of researchers were discussing the population of snow leopards in China, the country with the largest snow leopard habitat. The side event led to the creation of a network of Chinese researchers and organizations working on the snow leopard. This meeting of around forty young men and women (at least half were women) happened in the attic of a Tibetan restaurant on a side street in a slightly secluded part of the city. At one point I saw them pass around small chits of paper. A good friend who helped interpret for me said that they are going to vote on who would be the presiding person of the network. I was surprised. Voting was a taboo word in China. He told me that the government representatives in the room had authorized it because they had approved the candidates in the first place. The voting was to pick a leader from the government-approved list. On our way out I spoke to the restaurant owner, a very kind Tibetan man, in my broken Spitian and Ladakhi. I was naively surprised by the amount of attention I drew to myself after that. As I discovered, an Indian speaking in Tibetan to a Tibetan on the plateau aroused suspicion.

Since I was in China, my friends and colleagues from Peking University had arranged for me to attend the bigger event on the creation of the network of national parks. There, George Schaller gave a rousing talk about the biodiversity of the Tibetan plateau and he ended by thumping the table and questioning the authorities about the need for state-sponsored industrial-scale pika poisoning on the plateau which has been going on for over half a century. The plateau pika (*Ochotona curzoniae*), a rat-sized relative of the hare, is abundant on the Tibetan plateau and is perceived as a pest by the Chinese administration. Despite compelling evidence from George Schaller and many local scientists showing that pikas are important ecosystem engineers and beneficial for the health of pastures on the Tibetan plateau, the administration continues to pursue a policy of mass poisoning in an effort to eradicate them, with disastrous consequences for all local biodiversity. Schaller is the only Western scientist who has had extensive access to wildlife research on the Qinghai–Tibet plateau despite his confrontation of the government's policies.

'There would be serious consequences for such behaviour for anybody other than him,' my Chinese colleague whispered in my ear. I wondered if she was warning me or reminding herself.

That night, a handful of us walked with Schaller to a nearby restaurant for dinner. I matched my pace with Schaller's to ask him about the gorillas and the Virunga mountains. The rest of the group was walking ahead and he dropped his pace to answer me.

'Not many people ask me about *The Year of the Gorilla* any more,' he said.

As we walked through this new city in Xining, filled with glittering lights and billboards, we could feel a light chill in the thin, high-elevation air. Schaller spoke about the wet forests of Virunga, climbing the eight peaks of the mountain range and the human-like apes. We talked about Dian Fossey wanting to 'out-Schaller Schaller' when she went to study the mountain gorillas after him. I told him how I had read *The Year of the Gorilla* in the Aurangabad public library and decided to become a wildlife biologist. I thanked him for it and he shook my hand as we entered the restaurant where everyone was already discussing the menu. I didn't have the chance to speak to him about the snow leopards.

Two years before this second meeting with Schaller, on his and Matthiessen's trail with Gopal to the Shey Gompa, I was as keen as always to see a snow leopard. Gopal reminded me that the region was not like Spiti and that seeing a snow leopard in Dolpa is much harder. There are no researchers or spotters keeping track of snow leopard movement in the region. Schaller and Matthiesssen's accounts from forty years ago were the best available information. While that was true, the quest to see a snow leopard added a higher purpose to the expedition and if we succeeded against the odds, there would be an even greater sense of fulfilment.

My chain of thoughts was broken by the sweet melody of a bird singing from a bush. I stood still, waiting. A black and white bird, smaller than a pigeon, hopped out on a rock before hopping into another bush, pretending to be uninterested in me. The sight of the pied thrush reminded me of home. This bird would fly over Bangalore on his way to his wintering grounds in Sri Lanka, even stopping for a few days on the hills surrounding the town during its passage. I tried not to think about what my daughter would be doing at this time. Her third birthday was coming up in a few weeks. I had told myself that she was only 3 years old and wouldn't remember if I failed to make it back home in time. But I realized that I would never forget it or forgive myself if I did.

The day my daughter was born was the happiest I had ever been. Bhagya and I could not have foreseen how joyous it would be to become parents and to start a family together. We had to make some small and large adjustments, but parenthood changed us in profound ways and tinkering with our daily lives was a small part of the bargain. I lost my urge to climb mountains, and I lost my nerve to hold steady on steep precipices. For the first time ever, I thought about the consequences of my death; not in a morbid way, but in a calm, collected and calculated way. I knew it was time to stop climbing the more risky mountains.

After two days of walking we were about to reach Ringmo on the banks of the Phocksundo Lake. The path was steep but we covered it fast. My backpack felt heavy and the air in my lungs thin. I stayed hydrated and sweated profusely on that sunny day as we emerged from the birch forest onto the alpine meadows. We were now in snow leopard habitat. We climbed over the hill that had fallen in an earthquake 40,000 years ago and dammed the river to create Phocksundo Lake. The lake was shielded from us by other peaks, but we had climbed above the Suli gorge and could see the mountains and the Kang La Pass that separated us from Shey monastery on the other side.

Ringmo is a small village that sits on the edge of the lake where the water spills over a rocky bund and continues its journey down the stream through the Suli gorge to the Ganga. We walked into the village on a quiet afternoon. No one was around. We found our lodge and left our rucksacks in a small room in the wooden house. Our host was busy and did not stay to talk. Gopal was immediately occupied, making enquiries for his research and about the state of the Kang La Pass which we had to climb tomorrow. I walked to the banks of the lake to stare at the mountains and turquoise-blue water.

Somewhere across the lake was the place where Schaller had seen his snow leopard. Matthiessen had seen many animals but missed the snow leopard sighting. So far, I had seen nothing other than small birds. I had lost two days because I had refused to get in the car with the drunk drivers. The airfield at Dunai was not yet ready so I would need four extra days to walk back to Jajarkot to get a bus back to the border with India. I would add six days if I were to go up to Shey Gompa, since I needed two days to cross the Kang La Pass, two days to stay at Shey monastery and two more to return over the Kang La Pass to Phocksundo. I would add ten days to the trip if I decided to continue my journey.

There was no way to send a message home to my wife about the delays. She was alone with our daughter Tara and would worry about me. Even if I had managed to relay a message through some of Gopal's friends, I could not be sure that it would reach her.

Gopal joined me on the banks of the lake and we sat for a few minutes without speaking, before discussing his research. He wanted to compare the number of snow leopards in the Upper Dolpa region on the north side of the Kang La Pass with the Lower Dolpa region, this place around the lake and the meadows along the Suli gorge. He would estimate the population of blue sheep and livestock and confirm if the findings of my dumpling study were applicable here, and whether the region with more blue sheep also had a higher number of snow leopards and if they also killed more livestock. The eagerness in his voice reminded me of a younger version of myself.

I pointed across the lake to the place where I believed Schaller had seen the snow leopard and asked Gopal if I was correct. He gave a vague reply and added that the interest in seeing snow leopards is so high that the richest man in Nepal, the country's first billionaire who owns Wai Wai noodles, was trying to build a resort in Ringmo village for tourists trying to see the snow leopard. I wondered if Matthiessen had thought about the effect his book would have in mythologizing the elusive cat.

'You are lucky that you have seen this place before it changes forever,' Gopal said.

I didn't respond. We can resist change only to a certain extent. Our experience of walking to this place had been profoundly different from that of Matthiessen's and Schaller's. Although, once inside the park, the differences had thinned out. Ringmo seemed stuck in time. The Bon monastery inside the village matched Matthiessen's description. The Suli River and the forest had survived as well. But we had not been so fortunate as to see the langur monkeys and the Asiatic black bear that Matthiessen had seen.

Gopal traced a line in the sky pointing to the path that leads up to the Kang La Pass. He wanted to take Schaller's route on the way back via Saldang and said I could take Matthiessen's route. 'Then we can say we repeated Schaller and Matthiessen's expedition,' he added.

Schaller and Matthiessen's journey felt too high a bar to live up to. And I was feeling emotional turmoil, about leaving my 2-year-old daughter at home, the poverty I had seen along the route here, and the effects of mass tourism on remote places like these.

It would be another two years before a picture of over-crowding on the summit ridge of Everest would take social media by storm and spark a debate about the number of people, many with very little experience, attempting to climb the highest mountain for glory. Eleven people died that year and an expert assessed that at least five of them died directly or indirectly due to the over-crowding. The crowding on Everest had been a debate among my climber friends for a few years already.

'To aim for the highest point is not the only way to climb a mountain, nor is a narrative of siege and assault the only way to write about one,' Robert Macfarlane writes in the introduction to Nan Shepherd's book *The Living Mountain*.

I remembered Rajender's words, 'Everest is for tourists'.

I had found my personal mountain in the form of the snow leopard. It was not an animal I wanted to conquer, but one that I wanted to understand. For Matthiessen too, seeing the snow leopard was not the summit of his ambitions. He was a pilgrim on a journey to know himself better and the snow leopard opened a window onto himself.

I could see that what was happening to Mount Everest was going to happen to the snow leopard. It was only a matter of time before overcrowding, garbage and accidents would profoundly shape the experience of seeing the snow leopard. People would come imagining a world described by Peter Matthiessen and find something devoid of meaning.

'I don't want to go any further.' The words had been stuck in my throat for a long time. Gopal did not seem surprised. He understood, he said. I knew I wouldn't get to see a snow leopard if I turned back now, but this was not the most important thing on my mind. For the first time in my life, I was homesick.

Gopal and I spoke late into the night about his research and preparations for his winter fieldwork in Ringmo and then at Shey, beyond the Kang La Pass. When we parted the next morning, I wished him luck in his preparations for the winter and in seeing a snow leopard. I had had my fair share of snow leopard sightings over the years, but Gopal's first sighting could be transformational in how he understood the mountains and the animal for the rest of his life.

My journey back to Nepalgunj was difficult. My troubles started once I caught the run-down van. First we had a flat tire, then stones rolled down the mountain with the speed of a bullet threatening to shatter the windshield. And finally the car swerved in a mudslide on the road and slipped and turned over in a rice field. Twenty passengers and a goat crawled out of the windows as if it were the most normal thing to do. Nobody panicked and nobody was hurt. There was no screaming, arguing or fighting. A few people stayed back, talking to the driver, but most started walking down the road to the next village. Stupefied, I walked with the group that seemed to know what they were doing. That night, I found myself on the roof of a bus driving on some of the most treacherous mountain roads I'd ever encountered. Then it started raining and the six or seven of us young men who were on the roof took cover under a tarpaulin sheet.

'It will protect us from the rain but not the stones that might come off the cliffs,' someone said in Nepali with a few words of English and Hindi thrown in for my benefit. I felt claustrophobic and so I poked my head out of the tarp. It was as dark outside as it was inside. The spattering of the rain on the tarp and the tin roof of the bus drowned out all other sound. In a flash of lightning I had a glimpse of the gorge, the depth of which I could not fathom. The wall of wet black granite rock on the other side of the bus penetrated the clouds that were pouring raindrops the size of my fist. The bus stopped in the darkness, words were spoken and everyone on the roof got down; the bus started to move and we walked behind it. I noticed that we were passing under overhanging rocks so low that we would have smashed our skulls had the driver forgotten about us on the roof. The inside of the bus was packed like sardines and so once we had passed through the tunnel I went back to the roof. The darkness made it easy to forget the dangers that we were exposed to. If this bus were to slip off the road, it would kill us all. If a rock even the size of a penny were to fall on the head of one of us sitting on the roof, that would be the end.

And yet, I felt peaceful. I knew to expect these conditions when travelling in the mountains and the risks had long been part of my calculations. My thoughts wandered as I considered what snow leopards did in such weather. Would they prefer to be in a cozy cave with their long scarf-like tails wrapped around their bodies for warmth or would they be out hunting where the weather would mask the sound of them approaching their prey?

Late at night the bus stopped at a small village and some people got down. The seven of us from the roof moved inside the bus. Inside was a stench of sweat mixed with vomit and dust, but we were offered seats for the sacrifice we had made earlier. Two days later, I made it back to Bangalore in time for my daughter's birthday.

I heard from Gopal a couple of weeks later. He was excited about doing his MSc thesis at Shey-Phocksundo National Park. He had made arrangements to spend the winter there. 'But did you see a snow leopard?' I asked jokingly, knowing he would have mentioned it first, had he seen one. 'I am not worrying about it just yet. I know I will some day if I keep working honestly in this landscape,' he said dispassionately.

In 1973, Schaller had guessed there to be six snow leopards in the Shey, Saldang and Phocksundo regions. Schaller had seen one snow leopard and the pug marks of another one, and Matthiessen was 'pleased' to know that two of them were breeding. In 2017, with the aid of seventy camera traps, Gopal identified fourteen individual snow leopards in an area perhaps a little larger than the one covered by Schaller and Matthiessen. While I had retreated to the plains of Bangalore, Gopal was still searching for his first snow leopard sighting in Shey-Phocksundo National Park.

Early snow leopard researchers barely ever saw the animals they were studying. Even now, I see the snow leopard only a fraction of the time compared to the observations of my colleagues on other species like the tiger and the elephant. But this is what makes studying the snow leopard so special.

In December 2023 I was in the United States giving lectures at multiple universities. I was most excited about speaking at the

Cape May Zoo. The zoo had been a long-time supporter of the Snow Leopard Trust and had funded us for many years. It was my first time talking about my work at a zoo.

The zoo's vet received me and my colleague, and we wandered through the cold and wind-swept park. Most of their animals had taken shelter indoors. On our behind-the-scenes tour we were having an intense conversation about animal welfare and I was leaning against a fence post. The vet suddenly squinted his eyes, held my gaze and lowered his voice, saying, 'Don't be startled when you look behind you. Turn around slowly.'

Through the fence, only inches away from me, was the large round face of a snow leopard, Baatar. He looked intently at me, curious to understand who I could be in this place where he only ever saw handlers and the vet. He held me in his stare. His eyes were deep turquoise around a dark pupil ringed with sulphuric yellow that covered the entire visible part of the eyeball. He did not blink or look at the others, and I was spellbound, unable to move or look away. Baatar's face had a magical aura. For a brief second, he looked at my colleague (who had many years of experience of working at zoos) and then caught me again with his stare as I was about to say something. I was so mesmerized that when I moved away from the fence, I felt as if I had been released from a hold and only then could the muscles in my body relax. Baatar stayed near us on the other side of the fence, following our conversation. He would look at the person who was talking and would stalk anyone who turned their back to his fence. The handler told us that Baatar had a partner, but she was shy like her wild cousins and rarely made an appearance.

Cape May Zoo had been highly successful in breeding snow leopards. The cubs born there had been moved to populate other zoos in the country. I knew that Baatar would go on to live a long life, as snow leopards in zoos are known to live as long as 20 years. Sunshine would have been lucky to have lived half of that. The oldest recorded wild snow leopards are not more than 13 years old. Females in captivity don't breed past the age of 15 and yet are known to live as long as 22.

Most big cat species, except for perhaps the cheetah, breed well in captivity, although if managed badly this can result in inbreeding and the accumulation of bad genes that affect their health, well-being and survival. I was curious about how the snow leopards in the zoos were faring. The vet said that the Species Survival Plan for snow leopards was mindful of the problem of inbreeding and yet there were problems such as eyelid coloboma in many captive-born snow leopards, a condition where cubs are born with holes in their eyelids. Problems of inbreeding often manifest in deformities of the eyes, teeth and face, as well as the poor survival rate of newborns.

'If you look carefully at Baatar, you will see that he has a hole in his left eyelid,' the vet said after a pause. Up close, Baatar had a very gentle face, without the ferocity that we associate with big cats. His large round eyes roved around curiously. His spotlessly clear fur and serpentine tail looked more like a plush toy than the snow leopards I had handled in Mongolia. I could not spot the hole in his eyelid as the thick fur covered it well. It was hard to say farewell to Baatar as I felt a connection with him even after the short time that I had spent with him.

After Cape May, I stayed for a week at Stanford, where one of my colleagues had started her PhD after working with us on snow leopards. After my lecture at the department of Biology, I met a scientist who was leading a multi-institutional project to study the genetic diversity of snow leopards. Scientists from around the world had contributed tissue or blood samples of snow leopards from the wild. The DNA in an animal's body not only tells us about the individual as it is today but its evolutionary past. The genetic research showed that snow leopards have very low genetic diversity, the lowest among all big cats, even lower than the cheetah. For context, the cheetah has so little diversity that when a group of researchers tried skin grafting on twelve unrelated individuals, they all accepted the new piece of skin as their own. In theory, low genetic diversity in a species means that it will have a hard time adapting to new environments, climates and diseases. Scientists are struggling to explain how the cheetahs have survived with so little genetic variation. And now we have discovered that snow leopards have even less genetic variation and are more like each other than any other big cat.

The cheetah population passed through a population bottleneck around 10,000 years ago. At this time, very few individuals survived and the cheetahs today are the descendants of those few. I thought perhaps something similar had happened to the snow leopard during the last ice age when the mountains were covered in glaciers and little of their habitat remained. But there is no evidence to support this hypothesis. The scientist at Stanford found instead that snow leopards have had chronically small populations for hundreds of thousands of years. 'How small and how chronically?' I asked. To understand this, I had to understand a concept from genomics called the 'effective population size'. This is the number of individuals that can pass on all the genetic variation from one generation to the next. Humans, for example, have a population of about 8 billion, but the effective human population in genetic terms is only about 20,000. That means all the genetic diversity of humanity alive today could be represented by 20,000 genetically unique people. Based on the current genetic structure of the snow leopard, the team was able to determine that there have never been more than 14,000 effective snow leopards on Earth at any one time in the last 300,000 years. Their genetic diversity further declined by half about 200,000 years ago and then to only about 2,500 to 4,000 effective snow leopards about 20,000 years ago. A similar analysis of tigers found at least 150,000 effective tigers historically, indicating genetic variation in wild snow leopards is much lower than their closest relative, the tiger. This is not to be confused with the actual population size of the species. That number could be several orders of magnitude higher, as we have seen in humans.

Modern genomic techniques can show us amazing details about our historical past, and our adaptations today, yet it cannot help us estimate the total population of snow leopards without extensive field studies.

While we were still struggling to get a reliable estimate of overall snow leopard populations in the field, modern genomics had revealed that a chronically small population of snow leopards had led to low genetic variation in the species. Evolutionarily speaking, once the snow leopards specialized for a life in the rugged mountains, the total habitat available was restricted to the high mountains of Asia, and large parts of it were covered in glaciers for much of the time. Snow leopards seem to have survived in low numbers with low genetic diversity through several millennia. But the real challenge was yet to come. Tigers went from an unimaginably large number to less than 4,000 in just the last couple of hundred years. Only 7 per cent of historical tiger habitats have tigers today. How would snow leopards, with such low genetic variation, cope when the onslaught of development, land use and climate change reached the high mountains? If snow leopard populations get isolated or fragmented, that would quickly lead to inbreeding because of a lack of inherent genetic diversity within the species.

A couple of days later, I was due to deliver a lecture at the University of California at Berkeley. Because of the pace at which scientific publishing works, it would be a year before the results of this collaborative study on snow leopard genetics would be verified by peers and become available for the scientific community to read. There was a busy question-and-answer session and a lively after-talk drinks event on campus. But at times I struggled to stay focused. I felt envious of scientists using cutting-edge genomics who could say so much about the snow leopard based on a few small pieces of tissue, a few drops of blood or a few strands of hair. They didn't need to climb a mountain or even see a snow leopard in the wild. I had used genetics for a part of my PhD research when estimating the snow leopard population for specific small areas, but the tools available today are light years ahead. True, the genetic science of snow leopards was based on samples collected by more than twenty scientists climbing several mountains around the world. And the motivation behind my lifelong research was so that I could climb mountains and see snow leopards. This experience has given me a different and valuable perspective on the conservation of snow leopards. Genetics and fieldwork are both necessary in helping snow leopards survive.

I have been fortuitous in my search for the snow leopard. While

Schaller and Matthiesson had to penetrate the world of the snow leopard at Shey monastery after the monsoon rains and leave the region before the winter closed connections to the outside world, I had been able to enter this world in Spiti in 2008 at the beginning of the winter and lock myself in. Over a decade I had been able to learn what the place, its people, its animals and its mountains had to teach me. I was armed with enough training to summit Everest if I wished, but I was channelling my expertise to discover all I could about the snow leopard. In the words of my mountaineering teacher, Rajender, I had found my mountain, and it was the snow leopard. My goal was not to conquer it, but to understand it and, if possible, contribute to its survival. I was lucky to have lived among the people who have dwelled with the snow leopard for generations. They did not treat the snow leopard as an object to gawk at, just as they did not see the mountains that surrounded them as something to conquer.

Thinley, Prasen and I stood watching the tourists watching the Mother and the Siblings. Prasen noticed the melancholic look on my face and asked what I thought of the spectacle in front of us. The gentle smile on Thinley's face suggested that he had read my mind. Seeing a snow leopard in this milieu had little meaning besides a few likes on social media. It lacked the power to transform you as a person. It could not open a window into oneself or to any external hidden wonder. You learned nothing, felt no emotion; it was hollow. But I was happy for the money it brought to the local people. Tourism turned snow leopards from a threat to local livelihoods into a source of new cash income. This money had the power to convert the staunchest opponents into supporters of carnivore conservation. Nevertheless I worried about a corporate takeover of this industry. I would not be able to see the proud people of Kibber and Tashigang being reduced to menials. The kind of investment that a billionaire was proposing in Dolpa would be a disaster for snow leopards and the people who live alongside them.

Just then, somebody hugged me from behind and lifted me clean off the ground. When he let me go, I turned around to find the schoolteacher of Tashigang, smiling from ear to ear.

'It all began here fifteen years ago, remember?'

'How can I forget?' I said, returning his hug.

Thinley, Prasen, the teacher and I stood retelling the stories of our fieldwork in Tashigang, going skiing together, watching blue sheep and the early sightings of snow leopards.

'That may have been the beginning, but it was Rancho who created a watershed moment in the snow leopard tourism circuits,' Sushil said, as he joined the conversation. Sushil was now working for a tourism outfit.

A couple of years before the pandemic, on a bright sunny December morning in 2018, a male snow leopard had walked near to Kibber and sat down on a rocky ledge in full view of the village. Twenty years earlier, that would have meant his death by a herder protecting his livestock. Years of conservation efforts had helped increase snow leopard populations while also enabling the local people to bear the economic cost of living with large carnivores. The handful of tourists in the village that day were elated to see a snow leopard at such close quarters. When the snow leopard was seen again the next day, the sighting was captured on the tourists' phones and the news spread around the country. People from far-away places like New Delhi and Bangalore started planning a trip to Kibber. Tourists cancelled their tiger safaris in Ranthambore and headed to the mountains. Busloads of people began arriving.

Hardly anyone reporting on the sighting noticed that the snow leopard was old and unwell. Unable to move and unable to hunt, he was withering away each day. He moved slowly, although without a noticeable limp. None of this escaped the sharp-eyed people of Kibber, some of whom tried to feed him, leaving chunks of meat on the rocky ledge that he had made his last home. They named the snow leopard Rancho after the lead character of the Bollywood movie *3 Idiots*. In the movie, Aamir Khan, the actor playing Rancho, is a carefree and naturally brilliant engineering student who pursues an unusual career path, avoiding mainstream corporate jobs, nonetheless excelling as a scientist, inventor, teacher and a good human being.

Most snow leopards, like most carnivores, die a violent death.

Some die in territorial fights, or of injuries sustained while hunting on precipitous mountain slopes, and some die at the hands of humans who are either trying to protect their livestock or who want to trade in snow leopard body parts for money. But Rancho died peacefully. The people of Kibber looked after him during his old age. One sunny morning when Rancho did not move, the people of Kibber knew that he had passed away. They brought his body to the village square and gave him a funeral befitting a member of their own community. The people of Kibber felt a connection to Rancho, and I cannot help but think that Rancho felt it too.

# 12

# Un-endangered

Bright chandeliers shone from the ceiling and plush carpets adorned the floor. This was the banquet hall of the Presidential Residence in Bishkek, Kyrgyzstan, a room so luxurious that it resembled a palace. That evening, in August 2017, it was being prepared for a very important meeting: the first of its kind, where top bureaucrats, international leaders and politicians were assembling to discuss snow leopards and their future. The president of Kyrgyzstan, the vice-president of Afghanistan and the environment minister of Pakistan were among the delegates from nine other countries (China, Russia, Mongolia, Kazakhstan, Uzbekistan, Bhutan, Nepal, Tajikistan and India) representing the nations where snow leopards dwell in the wild. The gathering included ecologists like me – usually found in trekking boots and baggy pants – who had donned suits and ties for the occasion. A few dirt-tinged trekking boots peeped out beneath trouser hems.

There was hope in the air. If the people in this room took snow leopard conservation seriously, a lot could be done to save the species and its habitat in the majestic mountains of Asia. But some of us were aware of dark clouds gathering beyond the horizon. I had learnt that the International Union for Conservation of Nature, which publishes the Red List of Threatened Species, had decided to move the snow leopard from its current threat category of 'endangered' to a category of less-threatened species called 'vulnerable'. Widely used by policymakers across the world, the Red List is a measure of a species' closeness to extinction. Species listed as 'criti-cally endangered' are one step away from being extinct; next comes 'endangered' and then 'vulnerable'. If a species is listed as one of these three, it is considered threatened. The two other categories on this spectrum are 'near threatened' and those labelled as 'least concern'. Ironically, treating species as of least concern often leads to their decline and brings them closer to extinction, as we've found with the extinction of the passenger pigeon of North America, and the 99 per cent decline in the populations of Gyps vultures in India.

The downlisting of snow leopards might have been a reason for celebration. After all, we don't want the species to be threatened. However, researchers were sceptical of the data that the IUCN as-sessment team had used to reach their decision and political reasons that might have been a factor. Snow leopard habitats around the high mountains of South and Central Asia are under assault from commercial interests. Downlisting the snow leopard would make it easier for destructive industries, such as mining companies, to op-erate within the high mountains. Imagine this scenario: if a mining company wanted a licence to extract minerals from within a snow leopard habitat, they could now say that no endangered species – as defined by the Red List – inhabited the proposed mining site and that the licence should be granted at the earliest opportunity.

Snow leopard habitats around the world were under relentless pressure from mining. Only a few years earlier in 2009, in eastern Kyrgyzstan, the Kumtor Gold Company had obtained a licence to establish an open-cast mine across nearly one hundred square kilometres within the Sarychat-Ertash Nature Reserve, the crown jewel of Kyrgyzstan's national reserves. This was not a company with a stellar environmental record. On 20 May 1998, a truck from the Kumtor company had fallen into a river above a Kyrgyz village named Barskoon, spilling 1.7 tonnes of sodium cyanide. Within a month of the incident, several people had died, over 5,000 were reported sick and pregnant women were advised to have abortions. The case was settled by a Kyrgyz court twenty-two years later, and the surviving complainants of Barskoon compensated with the equivalent of $5,700 USD each. Fast-forward to 2009, and the endangered snow leopard was the sole species around which local conservationists could campaign to stop the company. Buoyed by international pressure, their efforts paid off, and the Kyrgyz government cancelled the mining licence and restored the reserve to its pre-2009 status.

In the banquet hall in Bishkek, I was unsurprisingly worried that the imminent IUCN's decision would make it harder for herders, environmentalists and conservationists to fight and win such battles. Nonetheless, one month after the presidential summit, on 14 September 2017, the IUCN officially declared the downlisting of the snow leopard from endangered to vulnerable. The date of the assessment was given as 8 November 2016 – the IUCN, it appeared, had delayed the official announcement for ten months, perhaps waiting for the Bishkek summit to be over, so that there would be less possibility of the range countries unitedly protesting their decision at the summit.

In 2017, when the IUCN announced the downlisting, I had one more reason to be upset, as some of my own work had directly contributed to this debate.

I first heard about the IUCN Cat Specialist Group's intention to reassess the conservation status of the snow leopard on the Red List back in 2015. Charu told me about it while we were on a run through the woods on the campus of the agriculture university in Bangalore.

'It seems that the committee has already made up its mind to downlist snow leopards from endangered to vulnerable without considering any serious evidence,' he said. I later discovered that the IUCN Cat Specialist Group had reached out to three snow leopard biologists, two from the US and one from the UK, to lead the threat assessment for snow leopards. All three were employed by large conservation NGOs from these countries and all had around thirty years of experience working in the mountains of Asia. I found it surprising that the IUCN could not find a single scientist from one of the twelve snow leopard range countries to be part of this group, despite several nationals serving as members of its Cat Specialist Group. I learnt later that one or two biologists from Asia had been invited to the panel, but had stepped down because they felt that their inclusion was tokenistic and the panel had already made its decision; their voices were neither heard nor their arguments recorded and their resignations in protest were never made public.

The previous assessment of snow leopard status had been made in 2008. It had concluded that snow leopards were to be categorised as 'endangered' because they had suffered severe declines in many Central Asian countries in the post-Soviet era due to inadequate protection.

Charu asked me how quickly I could get to Sweden. 'There is a Frenchman in Sweden who models carnivore populations.' We needed him to come on board and help produce a report on the demography of snow leopards because the final decision on the IUCN category was likely to be based on a few technical details.

'As soon as they grant me a visa,' I said without thinking.

Flying to Stockholm at a moment's notice to meet some Frenchman to stop a committee of gentlemen in Geneva from making a grave blunder felt like a scene out of a spy movie. I was lucky to get a visa within three days of applying for it, and I left that very evening.

The Assistant Director for Science of the Snow Leopard Trust received me at the airport, and we drove straight to meet the renowned wildlife population biologist Dr Guillaume Chapron for dinner at a small, nondescript restaurant an hour's drive outside of Stockholm. Guillaume had published some of the leading research papers on the population dynamics of wolves and tigers and had made a brief foray into studying the demography of snow leopards.

Guillaume worked at the Grimsö Wildlife Research Station, a field station of the Swedish University of Agricultural Sciences located in the middle of a verdant conifer forest, just a few hours' drive from Stockholm. During my visit in mid April, fresh snow lay on the neatly planted rows of trees. The research station had a rustic feel to it. I stayed in a centrally heated log cabin next to the main office building within walking distance of a beautiful lake where swans and ducks floated serenely. The range of research at this station was very impressive. Scientists studied different species from common cranes to wolverines, on issues pertaining to wildlife disease, crop and livestock raiding, and the effects of hunting. I could see why Guillaume had chosen to be based here rather than one of the better-known university departments in one of the European cities.

At dinner that first night near Stockholm, Guillaume was not thrilled by our proposal. At first, he did not see the value of this work for snow leopard conservation on the ground. He was busy. In Sweden, his research on wolves was heating up an already volatile debate between landowners and hunters on the one side and the urban population on the other over the maximum number of wolves which should be allowed in Sweden. Sweden has a policy of shooting excess wolves when their populations cross a limit. Guillaume argued that this limit was informed by the tolerance levels of the landowners instead of ecological science. Guillaume leaned heavily towards increasing the limit and was never afraid of controversy. I had heard that he was receiving death threats for his work on wolves.

We needed him, or rather his blessings, for our snow leopard report. I had even brought him a gift of an orange shirt. I had been told that orange was his favourite colour – but it had to be the right shade. A dash darker than the Dutch orange, a shade lighter than the Hindu orange. My wife Bhagya, who was six months pregnant with our daughter, found the right orange from an ethnic Indian store. Maybe it helped, as something clicked after our hour-long conversation, and he agreed to come on board as long as I did the heavy lifting. The mission had been accomplished. The actual report would be a matter of detail that I was confident of handling. To celebrate, I called the barman and asked for a vodka martini, shaken, not stirred.

The IUCN uses one of five criteria to list a species in one of the five categories. Criteria A uses information on the decline in the population of a species over the previous ten years or three generations of the species. Criteria B uses information on the area of distribution of a species. Criteria E is the only one that forecasts the probability of a species going extinct in the next ten years or three generations. If that probability is more than 50 per cent, then it categorizes the species as critically endangered. This criterion requires measurable information not only on the biology of the species but also of the threats to the species to be able to make convincing predictions of extinction probability. Criteria E was going to be nearly impossible to apply to a species like the snow leopard, which we know so little about.

The IUCN was not likely to apply Criteria A to snow leopards because there was little information on the species' actual decline in population. Since snow leopards are distributed over 3 million square kilometres, they would not fit Criteria B.

Therefore by elimination we knew that the IUCN committee would turn to either Criteria C or D to categorize the snow leopard, which delved into the number of mature individuals alive.

The question was simple: what proportion of the population of snow leopards could be considered sexually mature or 'mature individuals'? The IUCN required us to base our findings on an estimated snow leopard population of between 4,000 and 8,000 individuals – a starting premise which had no backing of evidence. According to Criteria C, if the answer was less than 2,500, then snow leopards would be categorized as 'endangered' and if this number was more than 2,500, then the snow leopards would be categorized as 'vulnerable'.

We built the basic structure of the model during the two weeks I spent in Sweden. At the end of these two weeks, both of us reached the same conclusion. According to the output of our models that relied on peer-reviewed data from the field, there was insufficient evidence to conclusively say that the number of mature individuals was more than 2,500. Unless there was conclusive evidence to prove that there were more than 2,500 sexually mature snow leopards, the downlisting of the species should not happen. I returned to India satisfied with my efforts.

Guillaume and I fine-tuned the model over email during the next few weeks and submitted our work as a report to my office. They would then submit the findings to the assessment team. I was not yet thirty and felt proud of myself. I had played my part in the conservation of the snow leopard and the high mountains of Asia. I was hopeful that the assessment team would change its mind once it saw the outputs of our models. We finalised our report on 3 June 2015, and I was planning to take a few days off from work.

After the completion of the report, my office suggested that I withdraw my name from the report. My employment at the Snow Leopard Trust – a research and conservation NGO (as opposed to a university or research institution) – might call into question my objectivity in building the model. A report submitted by a renowned French mathematician working from Sweden would be considered unbiased and more trustworthy. I was happy that I could be part of something bigger than myself, which mattered for the conservation of snow leopards globally, and was prepared to sacrifice my claim to authorship of the report for the greater good. This test I felt proved my unselfish commitment to the cause of snow leopard conservation. I agreed to withdraw my name, and the report was submitted to my office two days later.

A month later, my daughter was born. I did not have time to ask questions about racial power dynamics and the colonial legacies of conservation and research. I did not have the time to question the objectivity of the scientists who were leading the assessment for the IUCN; after all, they were employed by conservation NGOs like mine. I was swimming in the joy of parenthood and protected from sinking into the abyss of dark thoughts about neo-colonial structures of knowledge generation and conservation. And I believed that this report was going to save the snow leopards.

Come October, my optimism was dashed. In response to what was now called Chapron's report, three American authors had submitted a new paper using our model structure but concluding something very different. '[Guillaume] Chapron used his model to assess the proportion of mature individuals in the global snow leopard population. While his Leslie-matrix model is sound, we disagree with the definition he uses for mature individuals, which clearly contradicts that found in the Red List Guidelines (IUCN Standards and Petitions Subcommittee, 2014). In doing so, his model yields an erroneous estimate of the proportion of mature individuals and greatly overestimates the total population size needed to have 2,500 mature animals.'

It came down to semantics. The IUCN decision would be based on a narrow definition of 'mature individuals'.

The debate continued for a year and a half. I would hear murmurs about the definition of mature individuals during meetings, but I was never a direct part in the discussion. Then in September 2017, a little after the presidential summit in Bishkek, a committee led by four American and one British men, Tom, David, Peter, Rod and Kyle, concluded that snow leopard conservation had been successful, threat levels were reduced, and snow leopards should be downlisted from endangered to vulnerable.

At the presidential summit in Kyrgyzstan in 2017, when word had leaked about the impending move of the IUCN, the delegates from the governments of the twelve snow leopard range countries drafted a resolution protesting the IUCN's decision. I carried messages on little pieces of folded paper from the drafting team sitting in a small room and the main delegation in the meeting hall. Every word was scrutinized. For the first time, heads of states and foreign policies were entangled in the cause of snow leopard conservation. The draft was about to be ratified when China surprised everyone by backing out. Their delegation had received a message of disapproval from Beijing. The mood in the hall immediately changed. Heads hung low and angry words were spoken in whispers.

My colleagues who had experience with Chinese bureaucracy explained to me that China had found itself in an embarrassing position when they had questioned the IUCN over the downlisting of the giant panda in 2016. The IUCN had retorted that the Red List is prepared using a scientific process and not through national or international politics.

With China backing out, the protest statement could not be part of the official resolution of the meeting, famously called the Bishkek Declaration. But the Global Snow Leopard and Ecosystem Protection Program (GSLEP) secretariat, which organized the summit, put out a statement on its website on behalf of the other eleven countries. The strongest portion of the statement read, '[We] Reject categorically any change in the conservation status of snow leopard until PAWS [Population Assessment of the World's Snow leopards] generates scientific reliable population estimates and trends of the global Snow Leopard.'

The East versus West divide was clear. Large conservation NGOs from the West came out in support of the downlisting while regional country governments, local NGOs and the GSLEP secretariat questioned the move. Those opposed to the downlisting were accused of wanting to keep the snow leopard in the endangered category to ensure continued funding, while the big organizations favouring the decision were blamed for wanting to take the credit for having improved the status of a charismatic species in the Global South. Scientists and conservationists from both camps wrote opinion pieces in journals and gave media interviews clarifying their positions. Having agreed to withdraw my name from the report, I had conceded the little agency I had to be part of these discussions. During these arguments I was a forgotten pawn on the other side of the board.

The IUCN decision had been made anticipating criticism; three of the five members of the committee wrote in an article in a prominent conservation journal, 'Unbiased estimates of population density and abundance that incorporate uncertainty are essential components of any assessment of a species' status, including those for the Red List.' 'Estimates of population density and abundance' was precisely what was unavailable and yet they had found it not only apt to conduct an assessment but to also downgrade a species susceptible to an onslaught of global economic pressures.

They continued by saying, 'Downlisting on the IUCN Red List indicates that the species concerned is further from extinction, and is always to be welcomed, whether resulting from successful conservation intervention or improved knowledge of status and trends. Celebrating success is important to reinforce the message that conservation works, and to incentivize donors.'

In response to their article, Charu and Som Ale, a biologist from Nepal who is based in the West, wrote in the journal *Science* about the definition of 'mature individuals'. 'Records show just 3 of 344 captive leopards breed at 2 years old. Yet, the IUCN assessment used age 2 to 3 as the age of maturity'. They concluded, 'Desk-based announcements and celebrations of reduced extinction risks should be rejected in favour of rigorous field-based scientific evidence.'

Globally, the number of people involved in snow leopard research and conservation is very small. Almost everyone knows everyone else. Eventually both camps came together after the Bishkek summit and met in a classroom at the American University of Central Asia. The classroom setting was fitting because everybody needed some schooling. We all agreed on the need for more field data to conduct a better assessment.

The GSLEP secretariat proposed a bold plan to survey at least 20 per cent of the estimated 3 million square kilometres global range of the snow leopard by 2022. The survey would give a better estimate of the global snow leopard population using scientifically robust methods following a standardized protocol for field surveys and analysis, which all snow leopard scientists would agree to. A cool acronym was adopted, PAWS – Population Assessment of the World's Snow leopards.

I had visions of that snowy day ten years ago when I stood under that rock and imagined contributing to such an estimation. The dream was now coming true through the most unlikely route of a political controversy in the science of conservation.

Counting all the world's snow leopards is easier said than done. Snow leopards live in some of the highest mountain ranges of the world. They are distributed in some of the most rugged lands over roughly 3 million square kilometres, six times the size of France, and completely covered with mountains that are nearly twice the height of the Alps. However, this lofty challenge bridged the divide created within the snow leopard research and conservation community through the IUCN Red List assessment.

I had a hunch that everything we knew about snow leopards came from a handful of locations around the world. Places like the Spiti Valley in India, the Tost mountains in Mongolia and the Sarychat Nature Reserve in Kyrgyzstan. These were places that had a particular history of research, suitable access and infrastructure, and, most importantly, a healthy population of snow leopards on which scientists could record enough data to conduct their research. I undertook a research project to review the published studies on snow leopard population estimation. Munib, Charu and I were surprised to find that less than 1 per cent of the presumed distribution range of the snow leopard had been surveyed using standard scientific methods. Even these studies were biased in that they sampled relatively small areas in places that have a locally high abundance of snow leopards compared to the surrounding larger habitat expanses. We found a negative relationship between the size of the study area and the estimate of snow leopard density; meaning, studies with large study areas came out with lower estimates of snow leopard density and studies with small study areas found high snow leopard density. When you think of studying one of the most secretive and rare carnivores in some of the harshest terrains in the world, scientists go to the places where they have the best chances of finding these elusive cats. Extrapolating results obtained from the best possible habitats to the entire distribution range was not a good idea. The 20 per cent distribution range that PAWS hoped to survey could not only be along the known best habitats. It had to be representative of the entire region.

Around the same time, Örjan and his team tested scientists' ability to tell individual snow leopards apart by placing camera traps inside zoos and taking pictures of known snow leopards, then asking experienced and inexperienced researchers to identify the snow leopards from their pictures. They found that even experienced researchers make mistakes one in ten times – enough to inflate the resulting population estimate of the study.

The estimated global population of a species is a sacrosanct number for policymakers, bureaucrats, politicians and conservationists. In 2019, the prime minister of India, Narendra Modi, had proudly announced India's tiger numbers at a global conference in New Delhi. 'With around 3,000 tigers, India has emerged as of one of the biggest and safest habitats for tigers in the world,' he said. The numbers are important to create and showcase a particular image of the country. Estimating and monitoring the population of a species is a political tool in managing public perception. These two studies by Örjan and me had called into question the basis of current global snow leopard population numbers and questioned the premise of the IUCN assessment.

Kyrgyzstan had emerged as the main destination for politicians and scientists to host conferences and meetings about snow leopards. It was centrally located in Asia, with neutral or positive relationships with all global and Asian powers like the United States, Russia, China, India and Pakistan. It had simple rules regarding visas, great infrastructure in its roads and access to nature reserves like the Alatoo, which is only a couple of hours from Bishkek city.

In 2018, government officials and scientists gathered at a meeting held on the banks of Lake Issyk Kul, the same place where I met the falconer with Kuban. After the setback of the presidential summit and the surrounding media furore, the twelve snow leopard range countries unanimously agreed to start efforts to estimate snow leopard populations across its distribution range. The GSLEP secretariat, which was appointed the nodal agency for PAWS, set up an advisory panel of ten experienced snow leopard scientists and statisticians. This time, the panel consisted of three people from the United States and one each from Afghanistan, China, India, Nepal, Pakistan, Italy and Scotland, although still only one woman among the ten. Two of my senior colleagues represented the Snow Leopard Trust and the Nature Conservation Foundation on the panel.

I saw redemption in this global effort to estimate the snow leopard population. Validation that I was not wrong in generating the report with Guillaume despite the counter-report which the IUCN had ultimately relied on. Something good had come out of all the protests and pushback from the range countries. We may not have been able to stop the downlisting of the snow leopard, but the launch of the PAWS project by the groups on both sides of the debate was an acknowledgement that the current assessment was based on limited, even biased, data. I was aware however that although geographical biases about the size of the study areas were easy to acknowledge, political and racial bias about the scientists sampling and interpreting the data remained unaddressed.

I came back to India with a conviction that our team could do something remarkable with this opportunity. Our review of the world's studies on snow leopard populations had found that less than 10,000 of the estimated 3 million square kilometres of the snow leopard distribution range had been surveyed by researchers using rigorous scientific methods. The total snow leopard habitat in the state of Himachal Pradesh alone was 25,000 square kilometres. The largest area to have been surveyed was a mere 1,500 square kilometres in the Gobi in Mongolia.

Ajay, Munib, Thinley and the rest of our team had witnessed the scientific debate from the sidelines. They were as eager as I was to join in and contribute. We decided to take the challenge head-on and estimate the snow leopard population for the state of Himachal Pradesh – an area two and a half times larger than all the previous surveys put together. An area the size of Belgium covered in some of the tallest mountains and deepest gorges in the world. The survey would test us to our limits. Not only our physical ability to walk such a large and mountainous area to collect the necessary data but our ability to think through technical problems of study design and analysis.

# 13

# Counting Cats

'Don't die on me, Ramesh,' I muttered as I sank my ice axe into the hard snow on the slopes of Mount Kanamo. Ramesh Gulve, my good friend and a climbing buddy ever since I started mountaineering at the age of 17, had been gone for over fifteen hours and had not returned. I was alone at advanced base camp on that day in 2009, with night coming on fast. I had started climbing at 8 p.m. to look for him.

I gave up when I reached the mountain's shoulder at 5,500 metres. The remaining 500 metres to the summit were an exposed ridge line and I would have seen his lights if he were up there. I was hoping and praying that Ramesh had dug himself a snow cave and taken shelter for the night. The evening before, I had read him the story of how Hermann Buhl, the pioneering Austrian climber who first summited Nanga Parbat, had survived a night in the death zone above 8,000 metres by digging a hole in the snow for shelter.

Ramesh was a stronger climber than me, but the Spiti mountains were my backyard. As I walked along the exposed ridges, the Milky Way shone bright in the moonless sky. I was higher than all the dust and pollution, higher than any feature other than the peak of Mount Kanamo. The sky met the horizon well below my line of sight, and when I switched off my headlamp a warm glow of starlight reflected off the snow at my feet. Several thousand feet below, the lights of distant villages illuminated the dark, like little constellations. As a canopy of stars enveloped me from above, for a moment it felt as if

I were suspended in space. It was painful to experience something so magical when desperately worried about one of my best friends' survival.

Ramesh had come to visit my field site in Spiti to help me collect the snow leopard scat samples. The idea was to extract DNA from these scats and estimate their population for this region by identifying each individual through their unique structure. Ramesh was also trying to persuade me to join him on an Everest expedition the following year, and had been hoping to put in some practice climbs together.

I did not find Ramesh that night.

The next day, Ramesh was brought down by a team of rescue climbers from the village, with frostbite on his toes, but unshaken and otherwise fine. On seeing me, he told me that the story of Hermann Buhl had saved his life. After losing his way when returning from the summit, he had dug a hole in the snow and spent the night there at 5,600 metres, only a little further up from where I had given up my search. When I reached my highest spot at midnight, Ramesh must have been fast asleep in his cave. I could have walked right past and not noticed him.

Then, with a big smile, he pulled a Ziplock bag from his backpack. In it was a huge pile of snow leopard scat. He had found it soon after he began his ascent, carrying it all the way to the summit and his cave. As long as he had it, he had believed he'd see me again.

For the next few months, I worked in a molecular genetics lab in Bangalore to extract snow leopard DNA. It was harder than I had imagined. While the samples were well preserved because of the cold Himalayan climate, their exposure to UV light at high altitudes had caused desiccation.

A year later, when I finally worked out that there were eight snow leopards in the small 350 square kilometre region of Kibber in Spiti, Ramesh died on Everest. He had gone up to 7,000 metres to acclimatize and had suffered a cardiac arrest induced by altitude sickness when he returned to the advanced base camp. I was supposed to have been on that expedition with him. I am not sure if I could have saved him or if I would have shared his fate. I was then engaged to Bhagya and she had cried when she heard that I wanted to join an expedition to Everest. She reminded me that I had found the mountain of my dreams in studying the snow leopard. 'Wasn't that enough?' she pleaded. With that, I decided to focus on my PhD instead. Perhaps that snow leopard scat saved me and claimed Ramesh.

Between 2005 and 2015, the methods for estimating snow leopard populations had improved by leaps and bounds. The simplest method was to use camera traps. This method had been standardized for estimating the populations of the more charismatic cousin of the snow leopard, the tiger. During these ten years, it had been adopted and modified for use on snow leopards. Even then, as Örjan and his team had shown, errors in the identification of snow leopards are compounded during the statistical analysis and have led to the estimates being higher than the real population.

Molecular biology identification methods had also been improving. I used this method to estimate snow leopard population in multiple places for my dumpling study. I would collect faeces and bring them back to the laboratory to extract the DNA of the snow leopard that had produced the scat and do a similar statistical analysis to estimate the total number of snow leopards for the area. The main problem was that often the faecal sample did not have enough DNA in it to arrive at a good identification of the snow leopard. I had had a 30 per cent success rate in getting research quality DNA from the scats that we had collected in the past.

As the 2018 Bishkek summit came to a close, the buzz was about how to estimate snow leopard populations at larger spatial scales. It was not practical to train armies of scientists with camera traps or scat collection kits to survey larger and larger areas, in order to cover the entire 3 million square kilometres of the snow leopard habitat in the high mountains of Asia. With all the existing efforts, less than 1 per cent of the habitat had been surveyed. A simple scaling approach would take decades to survey the entire habitat.

Our team and I had learnt a lot over the past ten years. We had estimated snow leopard populations for small areas of 400 to 500 square kilometres using genetics. We had also used camera traps to estimate snow leopard populations in larger areas of up to 1,000 square kilometres. But our plan of sampling the entire 25,000 square kilometres of Himachal Pradesh was a leap.

Given the amount of lab work required to extract DNA from snow leopard faeces, I was not confident about using the method on a large scale. The molecular genetics world was also undergoing a change in technology, adopting a new method that we would have to standardize before it could be applied to estimate snow leopard numbers, which would take too long. No one in our team was a trained geneticist and we would have struggled to implement this method well. On the other hand, some of the people in our team were among the most experienced in the world in camera trapping for snow leopards. So we decided in favour of using camera traps to estimate the snow leopard population of Himachal Pradesh.

Our review of past research had already shown that scientists can be biased in how they choose sites to study snow leopards. To avoid this pitfall we relied on a distribution survey of the snow leopard conducted using an occupancy survey, to estimate the relative probability that a given area in Himachal Pradesh will have snow leopards. Using this map we were able to divide the state into three types of regions: regions with a high probability of snow leopard occurrence, regions with low probability and regions which are known to have snow leopards from past literature but which could not be surveyed in the previous study. Nearly 50 per cent of the snow leopard habitat in the state had a high probability of snow leopard occurrence while 30 per cent showed low probability; the remaining 20 per cent was yet to be surveyed. To avoid bias, we decided to sample in proportion to these probabilities.

This study design became a case study in the GSLEP secretariat's guidelines on the PAWS project for the twelve range countries. With the Indian Ministry of Environment, Forest and Climate Change and the Wildlife Institute of India, we prepared guidelines on how to conduct a countrywide snow leopard population estimation. The then Minister of Environment and Forest, Prakash Javadekar, released these guidelines and called the document SPAI, Snow Leopard Population Assessment of India.

But we soon learned the hard way that it was one thing to design a study based on the maps of the Himalaya and quite another to implement it at elevations of over 3,000 to 6,000 metres.

In May 2019, I sat in the shade of a pine tree on the banks of the Baspa River in the Sangla-Chitkul Wildlife Sanctuary of Himachal Pradesh with our team, conducting camera trapping for snow leopards. Our goal here was to survey the entire snow leopard habitat on the east side of the Satluj River, bordered by Tibet in the north and Uttarakhand in the east.

We were poring over several sheets of well-worn maps filled with illegible notes along the margins. The skies were a clear blue, and the sun was sharp, but a chill had descended from the snowy peaks of the Himalaya. The team leader, a young MSc student named Abhirup, was updating me about the progress of the surveys in this region. The lull induced by the calm gurgling of Baspa made it hard for me to stay attentive. The rest of his team was snoozing nearby; it was a rest day for them. They had been climbing up and down these mountains for nearly a month.

To set a single camera, a team of two people would walk anywhere between 2–10 kilometres and had to climb elevations up to 1,000 metres and sometimes more. The teams would check the cameras once a month and repeat the hike to retrieve them at the end of two months. This work required the stamina of a mountaineer and the diligence of a chess player in placing every camera at the correct location. By the end, every member of the team would have climbed higher than the elevation of Mount Everest. We would place more than 300 such cameras across the entire state.

Scientific discoveries are often credited to individual scientists, but it is rarely the case that we work in isolation. We had divided the state into ten sections. We covered some of the most beautiful and difficult mountain areas across the state. I was guiding multiple teams. Devika Rathore led fieldwork in Bharmour, Bhaga and the Miyar Valley in Lahaul; Harman Kaur in the Chandra Valley; Munib Khanyari in Upper Kinnaur; Aditya Malgaonkar in Pin Valley, Tabo and Upper Spiti; and Abinand Reddy managed the vast amount of data that poured from each site. Each team was comprised of about ten people who would be out in the high mountains for over three to four months each time.

Sitting next to the Baspa River in 2019, we realized that the pace of the camera setting was slower than expected. The team leader had recently done an expedition to set up some cameras in areas closer to the border with the neighbouring state. They had to cross the Rupin Pass, which stands at 4,650 metres above sea level, but snow and hailstorms hit them hard. They escaped with their lives, but another trekking group climbing from the other side was not so lucky. Three trekkers lost their lives to hypothermia. We had divided the study site into grids of 4 by 4 kilometres. Our plan was to set one or two cameras in each of the grids. 'But a few grids in the southern parts are completely inaccessible. We cannot reach those cliffs without risking our lives,' Abhirup said.

There was another problem. The northern side of the study area, which was easier to access, was too close to the international border with China. We had permissions from the headquarters of the Indo-Tibetan Border Police (ITBP) – after all, this was a project of the state government and a central ministry – but officers on the ground were reluctant. I wanted to make a last-ditch effort to convince the officers. We made the two-hour hike from the end of the road to the border checkpost of the ITBP. The landscape was open, more rolling hills than steep slopes. The Baspa River flowed more gently on this relatively flat ground; there were willow thickets on the bank, but little else stood more than two feet above the ground. Spring hadn't arrived yet, and the vegetation from the previous summer looked dead and smothered by the winter snow. The sun had melted away the snow, but the chilly wind had prevented new greens from sprouting. The days were hot, but the nights were still freezing. On the way, we came across the fur of blue sheep spread over the path. The skin, bones and horns were nowhere to be seen. We tried to rebuild the story of the last moments of the blue sheep's life. We kept talking about possible death due to an avalanche or predation by a snow leopard. We guessed that the scavengers would have carried away the skin and bones. Silently, we hoped that it had not been hunted by people, at least not by the ones we were going to meet.

The checkpost officer in charge was nearly a foot taller than me with wide shoulders and a handle-bar moustache. He towered over all of us and some of his men were bigger than him. He welcomed us warmly and, after checking our permission letter, invited us to lunch at their makeshift mess. We politely declined, saying we had already eaten. He asked us about our plans for the day and our work in general. Now I needed to bring up the question of camera traps in the areas between the end of the road and the camp – the international border was still a few kilometres away. I mentioned our larger project of estimating the snow leopard population in the entire state of Himachal Pradesh. He quickly told me that there were no snow leopards in their areas. 'There are some wild sheep and a fox or two but no leo-pard.' He stressed the word with a tinge of Haryanvi to his Hindi. 'An occasional bear comes near the checkpost at the road's end but nothing this far up in the mountains. I don't suppose you are interested in counting bears, are you?' he added after a pause. The body language of the ITBP men surrounding us was tense and I knew this was not a conversation that they wanted us to be having. I changed the topic and asked about the temple near the checkpost which I suggested we could visit. I remembered that Buddhist and Hindu people from the local region worshipped at the temple once a year. It was an old temple decorated with horns of long-dead blue sheep. 'We had come to assess the suitability of the habitat for different species of wildlife. The habitat looks great but there does not seem to be much wildlife here,' I offered as a truce. I didn't ask the question I had come to ask, and he did not have to say no. The air around us felt more breezy as the tensions dissipated. We had our tea, visited the temple and retreated to Chitkul village. We decided to limit our surveys to the areas between the road's end in the north and the inaccessible cliffs in the south. There were still nearly 500 square kilometres of snow leopard habitat in the middle.

The next morning, I joined the team to help finish setting up the remaining camera traps. I was paired with Dhamal from Kibber. His real name is Tanzin Thuktan, but in the twelve years that I have known him, I have never heard anyone call him that. Dhamal in Hindi translates to 'a blast of fun'. I have heard different stories of how he got the moniker. Dhamal is great fun but he is nothing like a blast. He is tall and thin and one of the quietest people on our entire team. In an hour-long conversation, he only ever speaks a sentence or two, but those few remarks tend to be the showstoppers. We were to set up a camera on a spur that was visible from Chitkul across the Baspa River.

We crossed the Baspa River using a footbridge and started our upward hike. I had estimated that we might have to ascend anywhere between 700 to 1,000 metres before we found a suitable place to set a camera trap for snow leopards. We were walking through a lovely birch forest. The bark of the trees was peeling like paper. The legend goes that Valmiki wrote the first draft of the epic Ramayana on the paper-like bark of this tree. We climbed in silence; I stopped sometimes to watch the colourful birds that crossed our path. Occasionally, we had to cross small streams, but we could use ice bridges made of avalanche debris that still held firm. We climbed out of the forest and walked through alpine meadows. The first sign of a large mammal was a pile of faeces left behind by a brown bear. Dhamal and I looked at each other and nodded; potential danger may be lurking around the next bend on the trail. When it comes to setting cameras for snow leopards, Dhamal is one of the most experienced people on our team. He takes his craft very seriously. He has been doing this for ten years, and his enthusiasm has only increased over time. When I pointed to a potential spot for our camera, he said softly, 'This is a livestock path; our datacard will be filled with images of sheep and goats. Let us climb higher up to those rocks, away from this meadow.'

The leaves were turning orange and there was a chill in the crisp Himalayan breeze. We had been out in the field tirelessly working for the past two years. The couple of years before that had been filled with the emotional turmoil of boardroom meetings and reports about the status of the endangerment of the snow leopard. Meanwhile, Tara, my daughter, was growing up fast – she was already 4 years old. I kept telling myself that she would not remember these years when she was older and it was okay for me to take some time for the important work I was doing. Once, when she was a little more than a year old, I came back home from a few weeks of travel and she ran back into the house, thinking I was a stranger. I laughed it off, blaming my scruffy look, stubbly beard, uncut hair and rugged field clothes, but I knew it was a warning sign. I kept telling myself that I could manage the tightrope balance between fieldwork and family life, but I was discovering there was no such thing as a balance between the two.

The fieldwork for estimating snow leopard populations across large spatial scales was tiring. It was exhausting to be out in the mountains all the time. The constant state of high alert was wearing the team down. Two members of our team had been chased by brown bears. They had escaped by making themselves look large by spreading out their hands while also fluttering their rain jackets. It was the most scared they had ever been. Others had had close calls on steep slopes when they were traversing with top-heavy backpacks. Yet the evenings at the camps were filled with stories and laughter, camaraderie and friendships. I could sense that everyone was looking forward to the winter so that we could take a break. The plan was to resume fieldwork in spring at the last site, the Great Himalayan National Park – a UNESCO world heritage site and one of the toughest to sample because of its rapid elevation change and lack of roads.

The weather forecast for the rest of September and October was looking good. One team had completed its work, and another was wrapping up. I approached Munib and Ajay with the suggestion of combining the two teams and completing fieldwork in the Great Himalayan National Park right away and not waiting for spring. After more than a decade of doing fieldwork in the Himalaya, I placed great value in a bout of favourable conditions. Making use of these opportunities – when seen after the fact – is often mistaken for good luck.

There was some apprehension because Ajay and Munib had a better sense of the team's morale and levels of tiredness. We spoke to the team members and there were some meek protests but also a palpable excitement at working within the crown jewel of the western Himalaya.

The Great Himalayan National Park is over 2,000 square kilometres, including the sanctuaries that abut it on all sides. It starts low at about 1,500 metres above sea level and reaches over 6,000 metres within a few kilometres as the crow flies. The vegetation changes rapidly from broadleaf to pine to birch to rhododendron to alpine meadows. The streams here flow strong with gushing white water which is almost impossible to cross without a bridge. The forested areas have common leopards and Himalayan black bears and the higher, more open habitats have brown bears and snow leopards. The goat-like goral were common along the forested hillsides, and there were reports of donkey-like serows in riverine forests. The majestic Himalayan tahr kept to the open rocky slopes. Musk deer with their sabretooth-like canines were common in the shrubby forests of the rhododendron bushes and blue sheep were spread out over the open alpine meadows. This place was as good as it got.

A large team of twelve people went to the Great Himalayan National Park with three large metal trunks full of camera traps. They studied the maps and worked out the trails and identified the places where they were likely to find snow leopards and began spreading out, but right then the weather moved in and it rained like there was no end to the monsoon for the next few days. Our team was pinned down in the tents in certain places and in others they continued placing cameras. We were in mission mode. All plans were being made at a hostel near the park headquarters in Sai Ropa. Locations would be marked on a map and teams of two to four people would use GPS to find these spots to place cameras. Once a team entered the park, there would be no communication with them. They would be on their own until they returned. Sometimes they returned earlier than planned but other times they were delayed and the team leader spent sleepless nights worried about their safety.

I had one eye on the weather forecast for the big snow. Our fieldwork would continue into October and November. We wanted to be sure that our cameras would not get buried in the snow or destroyed by avalanches. Once the stipulated sixty days of camera-trapping was over, our teams moved in once again and collected the cameras within a week. Only a week later there was snowfall in the higher reaches, shutting down all the passes and the park for the winter.

As December cold descended, we packed up and returned home. The research team had grown to more than thirty members, including those who were helping with the analysis. I promised everyone that we would all gather and celebrate our achievement of having completed one of the largest exercises in collecting data on the population of snow leopards. The news at that time was reporting a virus in Wuhan City in China, and that some people were stuck on a cruise ship in quarantine somewhere in Japan.

After completion of the fieldwork, we focused our attention on the analysis. Our cameras across the state yielded hundreds of thousands of images of wildlife. Of course, non-wildlife images had crept in too. When the data cards reached our office in Bangalore, the first step was to delete the pictures of local people who had been inadvertently photographed by our cameras. We had met the headmen and people of every village in our study sites and promised them that we would not compromise their privacy in any way.

Manvi Sharma, the colleague who had helped decipher the long-term blue sheep data and the causes of cyclicity, and other researchers in the team led the task of sorting, curating and analysing this large number of snow leopard images. To deal with the possibility of us making mistakes when identifying snow leopard individuals from images, we decided to repeat the process three times, each with a different researcher. If there was a one in ten chance of making a mistake each time, then the chance that the same mistake would occur all three times was much lower. This was the best that could be done, considering there was no real way to know.

As the virus began spreading globally, we all retreated to our homes, and in March 2020, like the rest of the world, India announced lockdowns. Once it became clear that the lockdowns would be long-drawn, Ajay reminded me that it was a good decision to press on and complete the fieldwork in the Great Himalayan

National Park while we had the opportunity. Those outside the teams thought we were lucky. The pandemic would have otherwise delayed us by one or even two years.

While most of us were busy collecting data on the population of snow leopards, Deepshikha Sharma had joined our team and she was quietly handling on-the-ground conservation projects. She is from the state of Himachal Pradesh and arguably had more at stake in the work. Deepshikha sported a streak of silver hair near her forehead and exuded confidence in what was still a male-dominated field. That confidence took the government officers and local political leaders by surprise. She had steadily led the conversation efforts from our office in Bangalore while much of the team was busy hiking up mountains and setting camera traps for the past two years.

After eight months of grappling with the data, between the first two waves of the virus and the lockdowns, we were ready to present the findings to the Principal Chief Conservator of Forest, Archana Sharma, and the Head of Forest Forces, Dr Savita Sharma, and the senior officers of the Himachal Pradesh Forest Department at the boardroom in Shimla. The wood panelling on the walls, the solid wood round table, and ancient trophy heads of ibex and argali on the wall were reminiscent of colonial times. I felt that it was a good opportunity for my younger colleagues, Manvi and Deepshikha, to develop their leadership skills and present the results in Shimla. Manvi had led the technical team analysing the data and Deepshikha was leading our conservation team in the state. The officers waited patiently for two hours while Manvi and Deepshikha explained the nuances of the method and the challenges of the process. The faces in the room bore a look that said, 'But tell us the number?' Manvi laid the groundwork by saying that as part of any statistical analysis, there is uncertainty around the estimate. She stressed that we had developed an estimate and not a count of the snow leopards in the state. Finally she revealed our figures. 'There are between forty-four and seventy-six snow leopards in the state with the estimate being fifty-two.' With Manvi and Deepshikha leading our team and Dr Savita Sharma and Archana Sharma from the Himachal Pradesh Forest Department, here were four women presiding over the meeting to announce the results of the first large-scale snow leopard population estimation effort. They had made history.

There was a sense of joy in the room. Himachal Pradesh was the first state in the country to have employed a scientifically rigorous method to estimate its snow leopard population. It was also the first region in the world to have followed the methodological guidelines of the GSLEP secretariat to estimate snow leopard population. The number 'fifty-two' was carried by tens of newspapers around the country over the days that followed. The newspapers were eager to publish a number that was not a death toll.

Our estimate was much lower than the expert opinions which had pegged it at ninety. This guess had been an extrapolation from a few small-scale studies like my own effort at estimating the population in Spiti. Unsurprisingly, more than 80 per cent of the state's snow leopards were outside protected areas. There was no need for immediate alarm: this was not a decline, but a correction of the estimate.

While this was a big achievement for us and the state, our effort was unlikely to raise a blip on IUCN's radar. Yet we were hopeful that we had shown the way forward in the effort to estimate the global population of snow leopards.

It was a watershed moment for our team. We had worked together under challenging field conditions and trying circumstances on a large-scale project that none of us were sure about. The goal had been achieved and everyone started to move on with their lives. Harman left for a PhD position at Stanford, Abinand took up a PhD position in statistical ecology at the University of St Andrews, Abhirup took up a PhD position at the Indian Institute of Science Education and Research in Pune, and Manvi became an Assistant Professor at Ashoka University in Delhi.

And there were new beginnings: Munib finished his PhD at the University of Bristol and started leading a similar effort to estimate the snow leopard population in the Union Territory of Jammu and Kashmir. The central government had recently bifurcated the state into two union territories. Virtually nothing was known about the snow leopard distribution and population in Jammu and Kashmir. Munib used our experience and trained team members from Himachal Pradesh to lead the effort in Kashmir. Other agencies were leading similar efforts in other Himalayan states in India and many more agencies were doing the same in other countries.

Deepshikha had meanwhile built a cadre of local women from the snow leopard habitats of Himachal Pradesh into an all-women's team of camera-trap experts. They were not only conducting the fieldwork to monitor snow leopard populations but also analysing the images to curate the data. It was a one-of-a-kind effort not found anywhere else in the world. These were the same women who had started out with us many years ago making knitted handicraft products under the name Shen. Deepshikha had taken the women out to the field and connected them directly to snow leopard research. When the heads of state of the twenty largest economies in the world met for the G20 summit in Delhi in September 2023, a short film about Deepshikha and the women of Spiti was featured in the main auditorium.

The mountains and rivers of Himachal Pradesh had been my world for the past fifteen years. From the time I had set foot here as an intern, I had come back every year and spent several months there at a time. I had dreamed of climbing these mountains and studying the snow leopard and I had fulfilled those dreams and been part of the major effort to estimate the world population of snow leopards. Every encounter added new and beautiful layers to my knowledge of the snow leopard, its mountain habitat, the ungulates that occupy the steep slopes and the warm and loving people who live alongside them. I had played my part in a revolution of understanding. As if summiting Mount Everest, I felt euphoria about all that we had achieved.

Two years later, reports on snow leopard numbers from other Indian states started to come out. We found that there were methodological differences across the states. Not everyone had paid attention to Örjan's research findings that even experts make mistakes in identifying snow leopards from images, which had led to a problem of overestimation.

Mongolia released a report in 2021 with a preliminary estimate of 953. In 2022, Bhutan estimated that there were 134 snow leopards in the country. In 2024, India's Ministry for Environment, Forest and Climate Change added up the snow leopard numbers from all the states and released a report estimating the snow leopard population in India at 718 snow leopards. An estimation of a thousand snow leopards was made for the Sanjiangyuan region of the Tibet–Qinghai plateau in 2025. A global number was now within reach.

Snow leopard numbers were making newspaper headlines around range countries. It reminded me of the famous line from Douglas Adams's *The Hitchhiker's Guide to the Galaxy*, in which the answer to the question of 'life, the universe and everything' is forty-two. Suddenly, the perception of the administration towards this species began to change, as if having an estimate of the snow leopard population meant that we understood this beautiful creature and its mountain home. Politicians and media spoke with great confidence about the ecology of the snow leopard and its conservation needs. These numbers became the ultimate truth.

Soon, the bureaucracy in the Himachal Pradesh Forest Department changed once again. The new set of officers had not experienced the highs of the PAWS project's heydays. They did not like the answer. They felt that fifty-two was too little. It was lower than some of the other states and countries. They called Deepshikha for a meeting and handed her a letter suspending all permissions for our research. Once again, I was reminded of *The Hitchhiker's Guide to the Galaxy*: 'Well, that's bureaucracy for you,' Slartibartfast said after the Vogons blew Earth to pieces.

Despite the challenges of this bureaucracy, Deepshikha and her band of local women continue to monitor snow leopard populations in and around their own villages. Through diplomacy and hard work she restored our research permissions and continues the ongoing mission to understand snow leopards and to promote their conservation.

In February 2024, my daughter Tara accompanied me to the field for the first time. She was not yet 9 years old. As our plane flew over the Himalaya and approached Leh, she pointed out of the window and said, 'Look, Daddy, the mountains are peeking through the clouds.' I had never before been anxious about seeing a snow leopard, but now I was. It had always just happened. I had missed seeing one at Phocksundo but I had barely spent three or four days in snow leopard habitat, and that in the middle of the summer.

My wife was traveling for her work and the idea of Tara accompanying me in the field was initially one of necessity, but as we started packing for the trip I had the same feeling as the first time I went to the Himalaya for a research project. Tara's school had agreed to give her a ten-day break on the condition that she would keep a diary and share everything she learnt on the trip with the rest of her class. Tara was excited, more about the trip than the prospect of seeing a snow leopard. I was excited about the possibility of being able to show her a snow leopard.

As our flight descended into the Indus Valley towards the airport at Leh, the region was covered in a blanket of snow. The seatbelt could barely keep Tara from jumping up and down. I knew it would be a challenge to keep her calm for the first two days of acclimatization.

I was also nervous. What would she feel if we did not see a snow leopard? Would her friends make fun of her? I did not want to let her down. I had led the effort to estimate the population of the species for an entire state and yet I could not stop wondering whether I would be able to show her a snow leopard. More than fifteen years of asking questions about them, and I had learnt that snow leopards were like the mountains they inhabit – there is something unknowable about them.

The first day after acclimatization we drove out to a study site and did not see anything all day. Not a hare, an ibex or blue sheep. A golden eagle in the distance which I was very happy about did not even register a shadow of an impression on Tara. But she had a great time playing in the snow.

The next day I had meetings in town, so Karma took Tara to watch an ice hockey game between two regiments of the Indian Army.

Karma called that night while Tara was telling me about the ice hockey game. He said that a tourist outfit had seen a snow leopard in a nearby valley. There was a good chance that the snow leopard would still be around the next day. 'Let's go there tomorrow,' I said, cutting him off mid-sentence.

We had an early breakfast and I packed two boxes of snacks, a box of lunch and some chocolates for Tara. When we reached the place where Karma was expecting the snow leopard to be, Tara was surprisingly calm. I checked every rock and bush that poked out of the snow through my binoculars. We were in a wide valley with deep snow on one side but very little on the other. Strong winds and sun had cleared snow on the south-facing side of the valley. The stream flowing below was completely frozen over. The sky was a luminous blue with wisps of clouds. It was the perfect place and the perfect weather to show a kid a snow leopard.

We were waiting with guides and tourists from a couple of different outfits. We had picked a spot with black shale on the ground in a little pocket that was protected from the wind and had direct sunlight so it felt warmer than the surrounding area. Karma had the spotting scope and I kept Tara engaged by showing her the folds in the different sediment layers of the Himalaya. We spoke about the Himalaya emerging out of the Tethys Sea. Her teacher had told her about it, and she corrected me every time I deviated from her teacher's version. In the back of my mind I was wondering about what would happen if the snow leopard didn't materialize, and whether I'd be able to bring Tara to the Himalaya again. Would climate change's impact on the landscape make a sighting more difficult over the coming years? Would there even be snow?

Tara pointed at the different peaks and asked which ones I had climbed. Just then, there was a buzz in the air. Feet scrambled, the tone of the whispered conversations changed, and every muscle in every human body tensed. Karma put a hand on my shoulder and said that the snow leopard had stepped out into the open. He was heading towards a sunny spot on a rocky ledge on the ridge across from us. Karma continued to follow the snow leopard with the spotting scope. I used my binoculars and trained them on the general area and picked up the snow leopard's silhouette on the snow. Tara was patient and waited.

When the snow leopard, a beautiful male with thick grey fur covered in black rose petals, lay down on a rock slab a little higher than us, I picked up Tara to help her reach the eye-piece of the spotting scope mounted on a tall tripod. I had to be the one showing her the snow leopard at her first sighting.

'Daddy, I see him. He is looking back at me.' Tara spoke in a high-pitched, excited yet restrained voice, without taking her eyes off the spotting scope.

Two air force fighter jets made a low pass over the valley, oblivious to the people and the snow leopard below. The snow leopard and Tara both looked up at the loud and strange sound coming from the sky.

'Now you take a look, I want to try and spot him through my binoculars,' she said, wriggling out of my grip and running down the icy road.

The snow leopard stayed on the ridge a long time and Tara watched him throughout using her tiny pair of binoculars. She wanted to see the snow leopard 'by herself'.

Soon it was time for Tara to have lunch. There were many distractions but the snow leopard helped. On muscular legs he walked down to a thicket of willow bushes where he had hidden his kill from a few days ago. It was a large blue sheep, hard to see through the dried woody tangle, but Tara gave me a wide-eyed look when she saw him pull out a pink ribbon of flesh.

'It's time for lunch. Even the snow leopard is taking a break to have his meal.'

After several more hours of watching the winter sun moved lower on the horizon as cold air descended from the surrounding peaks into the valley. Tara huddled next to me for warmth and drew the snow leopard's curved tail on the frozen ground with her feet. This was the Eden described by Peter Matthiesen all those years ago. The snow leopard, a talisman to so many, had been my guiding light for eighteen extraordinary years, during which I had experienced many professional and personal adventures. The snow leopard got up and stretched, causing another wave of excitement in Tara and me before he slunk off into the crags of the mountainside. I took a deep breath of the crisp mountain air. Tara raised her arms, indicating she wanted to be carried. She dozed off on my shoulder as the three of us turned back home.

# Notes

## Introduction

Schaller, G. (1972). 'On the behaviour of blue sheep (*Pseudois nayaur*)'. *Journal of the Bombay Natural History Society*, *69*(3), 523–37.

Schaller, G. (1971). 'Imperiled Phantom of Asian Peaks'. *National Geographic*, *140*(5), 702–7.

## Chapter 1

Callaghan, A. (5 June 2023) '29 People Died in One of the Worst Mountaineering Accidents in History. What Happened?' *Outside Magazine*. Retrieved from https://www.outsideonline.com/outdoor-adventure/exploration-survival/avalanche-mountaineering-accident-draupadi-ka-danda-2/ (accessed on 5 August 2025).

## Chapter 2

Hemmer, H. (2023). 'An intriguing find of an early Middle Pleistocene European snow leopard, *Panthera uncia pyrenaica* ssp. nov. (Mammalia, Carnivora, Felidae), from the Arago cave (Tautavel, Pyrénées-Orientales, France)'. *Palaeobiodiversity and Palaeoenvironments*, *103*(1), 207–20.

Lovari, S., & Ale, S. (2001). 'Are there multiple mating strategies in blue sheep?' *Behavioural Processes*, *53*(1-2), 131–5.

Suryawanshi, K., Bhatnagar, Y., & Mishra, C. (2010). 'Why should a grazer browse? Livestock impact on winter resource use by bharal *Pseudois nayaur*'. *Oecologia, 162*(2), 453–62.

## Chapter 3

Forsyth, D., & Hickling, G. (1997). 'An improved technique for indexing abundance of Himalayan thar'. *New Zealand Journal of Ecology, 21*(1), 97–101.

Mishra, C., Allen, P., McCarthy, T., Madhusudan, M., Bayarjargal, A., & Prins, H. (2003). 'The role of incentive programs in conserving the snow leopard *Uncia uncia*'. *Conservation Biology, 17*(6), 1512–20.

Mishra, C., Van Wieren, S., Ketner, P., Heitkönig, I., & Prins, H. (2004). 'Competition between domestic livestock and wild bharal *Pseudois nayaur* in the Indian Trans-Himalaya'. *Journal of Applied Ecology, 41*(2), 344–54.

Suryawanshi, K., Bhatnagar, Y., & Mishra, C. (2012). 'Standardizing the double-observer survey method for estimating mountain ungulate prey of the endangered snow leopard'. *Oecologia, 169*(3), 581–90.

Suryawanshi, K., Redpath, S., Bhatnagar, Y., Ramakrishnan, U., Chaturvedi, V., Smout, S., & Mishra, C. (2017). 'Impact of wild prey availability on livestock predation by snow leopards'. *Royal Society Open Science, 4*(6), 170026.

## Chapter 4

Suryawanshi, K., Bhatia, S., Bhatnagar, Y., Redpath, S., & Mishra, C. (2014). 'Multiscale factors affecting human attitudes toward snow leopards and wolves'. *Conservation Biology, 28*(6), 1657–66.

Bjerke, T., Kaltenborn, B., & Thrane, C. (2001). 'Sociodemographic correlates of fear-related attitudes toward the wolf (*Canis lupus lupus*). A survey in southeastern Norway'. *Fauna Norvegica, 21*, 25–33.

## Chapter 5

Johansson, Ö., Rauset, G., Samelius, G., McCarthy, T., Andrén, H., Tumursukh, L., & Mishra, C. (2016). 'Land sharing is essential for snow leopard conservation'. *Biological Conservation*, *203*, 1–7.

## Chapter 6

Berger, J., Buuveibaatar, B., & Mishra, C. (2013). 'Globalization of the cashmere market and the decline of large mammals in Central Asia'. *Conservation Biology*, *27*(4), 679–89.

Samelius, G., Suryawanshi, K., Frank, J., Agvaantseren, B., Baasandamba, E., Mijiddorj, T., … & Mishra, C. (2021). 'Keeping predators out: testing fences to reduce livestock depredation at night-time corrals'. *Oryx*, *55*(3), 466–72.

Lkhagvajav, P., Alexander, J., Byambasuren, C., Johansson, Ö., Sharma, K., Mishra, C., & Samelius, G. (2024). 'Snow leopards and water: high waterhole visitation rate by a breeding female snow leopard in summer'. *Snow Leopard Reports*, *3*(1), 41–45.

Nyam, E., Alexander, J., Byambasuren, C., Johansson, Ö., Samelius, G., & Lkhagvajav, P. (2024). 'Snow Leopard digging for water in an arid environment'. *Snow Leopard Reports*, *3*(1), 37–40.

## Chapter 7

Arias, M., Coals, P., Ardiantiono, Elves-Powell, J., Rizzolo, J., Ghoddousi, A., Boron, V., da Silva, M., Naude, V., Williams, V. Poudel, S., … & Dickman, A. (2024). 'Reflecting on the role of human–felid conflict and local use in big cat trade'. *Conservation Science and Practice*, *6*(1), e13030.

Bijoor, A., Khanyari, M., Dorjay, R., Lobzang, S., & Suryawanshi, K. (2021). 'A need for context-based conservation: incorporating local knowledge to mitigate livestock predation by large carnivores'. *Frontiers in Conservation Science*, *2*, 766086.

Nowell, K., Li, J., Paltsyn, M., & Sharma, R. (2016). 'An ounce of prevention: snow leopard crime revisited'. TRAFFIC, Cambridge, UK. Available from: https://www.traffic.org/site/assets/files/2358/ounce-of-prevention.pdf (accessed 5 August 2025).

O'Connor, V., Thomas, P., Chodorow, M., & Borrego, N. (2022). 'Exploring innovative problem-solving in African lions (*Panthera leo*) and snow leopards (*Panthera uncia*)'. *Behavioural Processes*, 199, 104648.

## Chapter 8

Kapadia, H. (1999). *Across Peaks and Passes in Himachal Pradesh*. Indus Publishing Company, New Delhi.

Khanyari, M., Zhumabai uulu, K., Luecke, S., Mishra, C., & Suryawanshi, K. (2021). 'Understanding population baselines: status of mountain ungulate populations in the Central Tien Shan Mountains, Kyrgyzstan'. *Mammalia*, 85(1), 16–23.

## Chapter 9

Paine, R. T. (1980). 'Food webs: linkage, interaction strength and community infrastructure'. *Journal of Animal Ecology*, 49(3), 667–85.

Paine, R. T. (1969). 'A note on trophic complexity and community stability'. *The American Naturalist*, 103(929), 91–3.

Janecka, J. E., Hacker, C., Broderick, J., Pulugulla, S., Auron, P., Ringling, M., ... & Jackson, R. (2020). 'Noninvasive genetics and genomics shed light on the status, phylogeography, and evolution of the elusive snow leopard'. In J. Ortega & J. Maldonado (eds.), *Conservation Genetics in Mammals: Integrative Research Using Novel Approaches* (Cham and New York, Springer, 2020), pp. 83–120.

Adichie, C. (2009). *The danger of a single story* [Video]. TED Conferences. https://www.ted.com/talks/chimamanda_ngozi_adichie_the_danger_of_a_single_story (accessed 5 August 2025).

Sharma, M., Khanyari, M., Khara, A., Bijoor, A., Mishra, C., & Suryawanshi, K. (2024). 'Can livestock grazing dampen density-dependent fluctuations in wild herbivore populations?' *Journal of Applied Ecology*, *61*(6), 1243–54.

Elton, C., & Nicholson, M. (1942). 'The ten-year cycle in numbers of the lynx in Canada'. *The Journal of Animal Ecology*, *11*, 215–44.

Krebs, C., Boutin, S., Boonstra, R., Sinclair, A., Smith, J., Dale, M., ... & Turkington, R. (1995). 'Impact of food and predation on the snowshoe hare cycle'. *Science*, *269*(5227), 1112–15.

## Chapter 10

Patel, J., Sharma, M., Khanyari, M., Bijoor, A., Mishra, C., Harihar, A., & Suryawanshi, K. (2024). 'Influence of predator suppression and prey availability on carnivore occurrence in western Himalaya'. *Journal of Zoology*, *322*(1), 3–11.

Patel, J., Khanyari, M., Malhotra, R., Pawar, U., Bijoor, A., Sharma, D., & Suryawanshi, K. (2024). 'Snow leopards or solar parks? Reconciling wildlife conservation and green energy development in the high Himalaya'. *Biological Conservation*, *299*, 110793.

## Chapter 11

Khanal, G., Mishra, C., & Suryawanshi, K. (2020). 'Relative influence of wild prey and livestock abundance on carnivore-caused livestock predation'. *Ecology and Evolution*, *10*(20), 11787–97.

Macfarlane, R. (2014). Introduction. In N. Shepherd, *The Living Mountain*. Canongate Books, Edinburgh, vii–xxxii.

Solari, K., Morgan, S., Poyarkov, A., Weckworth, B., Samelius, G., Sharma, K., ... & Petrov, D. (2023). 'Extreme in every way: exceeding low genetic diversity in snow leopards due to persistently small population size'. *bioRxiv*, 2023–12.

## Chapter 12

Bishkek Declaration 2017 (24–5 August 2017). https://www.cms.int/sites/default/files/declaration.pdf (accessed on 5 August 2025).

Johansson, Ö., Samelius, G., Wikberg, E., Chapron, G., Mishra, C., & Low, M. (2020). 'Identification errors in camera-trap studies result in systematic population overestimation'. *Scientific Reports*, *10*(1), 6393.

Suryawanshi, K. , Khanyari, M., Sharma, K., Lkhagvajav, P., & Mishra, C. (2019). 'Sampling bias in snow leopard population estimation studies'. *Population Ecology*, *61*(3), 268–76.

Mallon, D., & Jackson, R. (2017). 'A downlist is not a demotion: Red List status and reality'. *Oryx*, *51*(4), 605–9.

Ale, S., & Mishra, C. (2018). 'The snow leopard's questionable comeback'. *Science*, 359(6380), 1110.

McCarthy, T., Mallon, D., Jackson, R., Zahler, P. & McCarthy, K. (2017). *Panthera uncia. The IUCN Red List of Threatened Species* 2017: e.T22732A50664030. https://dx.doi.org/10.2305/IUCN.UK.2017-2.RLTS.T22732A50664030.en (accessed on 5 August 2025).

Chapron, G. (2015). 'Modelling the proportion of mature individuals in snow leopard populations'. Unpublished report.

## Chapter 13

Ghoshal, A., Bhatnagar, Y., Pandav, B., Sharma, K., Mishra, C., Raghunath, R., Suryawanshi, K. (2019) 'Assessing changes in the distribution of the Endangered snow leopard *Panthera uncia* and its wild prey over 2 decades in the Indian Himalaya through interview-based occupancy surveys'. *Oryx*, *53*(4), 620–32.

Suryawanshi, K., Reddy, A., Sharma, M., Khanyari, M., Bijoor, A., Rathore, D., . . . & Mishra, C. (2021). 'Estimating snow leopard and prey populations at large spatial scales'. *Ecological Solutions and Evidence*, *2*(4), 12115.

Ministry of Environment, Forest and Climate Change (2024).

*Status of Snow Leopard in India. The snow leopard population assessment in India (SPAI).* https://wii.gov.in/images/images/documents/publications/SPAI_report_2024.pdf (accessed on 5 August 2025).

# Acknowledgements

The idea for this book has been in my head since my first visit to the Himalaya, but it started taking shape through conversations with Suri Venkatachalam, and then Harini Nagendra helped me set the course towards making it a reality by encouraging me to apply to the Wissenschaftskolleg Zu Berlin (Wiko). My deepest thanks to you both.

Much of the proposal was developed at the Wiko, where I truly gained time to think. I am grateful to the Dean of Wiko, Daniel Schönpflug, for discussions and guidance. At Wiko, my thanks to my fellow Fellows and colleagues Jessica Metcalf, Sean McMahon, Alyx Cullen, Ilya Kliger, Britt Koskella, Mike Boots, Chris Kelty, Hannah Landecker, Guy Tillim and many others. You inspired me and gave me the confidence to do this. My time at Wiko was made even more rewarding thanks to Eduardo Halfon and Lucia Corral, and the countless hours of discussions with Eduardo helped me hone the craft.

I am grateful to the Canadian Institute For Advanced Research (CIFAR) and my colleagues in the Future Flourishing programme for the opportunity to discuss my ideas. CIFAR has helped me take new directions in my work on human-nature relationships.

My biggest thanks to my editors, Bella Lacey, Courtney Young and Manasi Subramaniam. I learned a lot from you and the book benefited immensely from your supervision. I had imagined this book differently until Sarah Rigby helped me take my reader's perspective, and as cliched as it may sound, I don't have the words to thank Sarah. I am also grateful to Sarah for helping me find Patrick Walsh and PEW Literary. Patrick has been by my side at every step of the way. Thank you to Cora, Margaret, Alex and others in PEW Literary who have been a huge help all along.

I am grateful to my academic mentors, Charudutt Mishra, Yash Veer Bhatnagar and Steve Redpath, you have all helped me in my career as a snow leopard biologist. Charu taught me the rigour of science, Yash Veer reminded me to stay in love with the mountains while still being objective in my research, and Steve inspired me to engage with the arts. This book is my attempt at blending the three.

At the Nature Conservation Foundation, I am indebted to T. R. Shankar Raman, Divya Mudappa and Rohan Arthur who took an interest in everything I wrote and were available for a conversation every time I needed someone to talk things over. Aparajita Datta, Suhel Quader, Anindya Sinha and M. D. Madhusudan supported me at every step of my career. Professor Mahesh Rangarajan always encouraged me to write for a wider audience and nudged me to think about a book. At the Snow Leopard Trust partner network, I am grateful to Koustubh Sharma, Siri Okamoto, Laura Farnitano, Brad Rutherford, Marissa Niranjan, Justine Shanti, Rakhee Karumbaya, Örjan Johansson, Gustaf Samelius, Bayarjargal Agvaatseren, Purevjav Lkhagvajav, Nadia Mijiddorj, Odbayar Tumendemberel, Lingyun Xiao, Cheng Chen, Juan Li, Tang Piao Piao, Kubanych Jumabay, Cholpon Abasova and many others who made the research and conservation projects possible. I will carry memories of Sumbee with me forever.

At the National Centre for Biological Sciences, a huge thanks to Jayashree Ratnam, Mahesh Sankaran and Uma Ramakrishnan for being friends, collaborators and advisors. Geoff Hyde, our language and writing tutor during the master's programme, made me believe that it was possible for me to write and publish in a foreign language like English. Ajith Kumar, the director of the master's program in Wildlife Biology and Conservation, you were my first teacher of ecology, and I owe you everything. My friends from the master's course Nandini, Umesh, Swapna, Priya, Deepti, Aathira, Divya, Kiran, Nachiket, Dharma, Priyanka, Kaavya, Rohini and Robin for encouraging audacious ideas. I am especially grateful to Nandini Velho; some of the ideas in the book were first written in and inspired by our letters, emails and postcards. Thanks to Aathira for helping with an early draft of the sample chapter.

The people in Kibber, Tashigang, Tost, Sarychat and Dolpa who hosted me in their homes this past decade and a half; I will never forget your warmth and love. Sushil, Padma, Thinley, Kalzang Gurmet, Chunit Kesang, Tandup Chhering Kamal, Tanzin Thukten Dhamal, Rinchen Tobge, Kalzang Pulzor, Thuktan Kalzang, Lalung Thukten, Takpa-ji, Lama Tanzin Chewang, Kibber Takpa, Karma Sonam, Rigzin Dorjay, Lobzang, Midji, Oyuna and all the others who have been part of my research and mountaineering teams. Colleagues, friends and students at the High Altitude Program, Ajay Bijoor, Rishi Sharma, Radhika Timbadia, Ranjini Murali, Saloni Bhatia, Munib Khanyari, Deepshikha Sharma, Dipti Humraskar, Deepti Bajaj, Devika Rathore, Gopal Khanal, Jenis Patel, Abhirup Khara, Harman Kaur, Abinand Reddy, Aditya Malgaonkar, Abhishek Ghoshal, Mayank Kohli and many others. I apologize to the many people whom I have had the privilege to work with but have not been able to name here and in the book.

Thank you to my uncle, Satish Ingle, for taking me on my first trek in the Sahyadri mountains. You sent me on a new trail in life, and I am eternally grateful. Huge thanks to Prasenjeet Yadav, who been like a brother-in-arms in everything I did. Thanks to my friends Karan Tambe, Dhanushree Tambe and Abhishek Chandhere for being supportive throughout. Thanks to my climbing buddies at NIM and HMI, Harsh Sapdhare, Kumar and Rajesh for the memorable times. Ramesh Gulve and Tsering left this world too early and you will always be in my thoughts.

Lastly, thank you to my family. My parents, Sunita and Tatya, were always supportive of my unconventional choices. My sister Neha has inspired me with her dedication and focus. My life partner Bhagyashree was my first reader, without whom I would have never finished writing the first draft. My daughter Tara, who inspires me every day.

# Index

[to come, 12pp available]